OPENING SWITCHES

ADVANCES IN PULSED POWER TECHNOLOGY

Series Editors:
A. Guenther, *Air Force Weapons Laboratory, Kirtland AFB, New Mexico*
M. Kristiansen, *Texas Tech University, Lubbock, Texas*

Volume 1   OPENING SWITCHES
              Edited by A. Guenther, M. Kristiansen, and T. Martin

---

A Continuation Order Plan is available for this series. A continuation order will bring delivery of each new volume immediately upon publication. Volumes are billed only upon actual shipment. For further information please contact the publisher.

OPENING SWITCHES

Edited by

A. Guenther
Air Force Weapons Laboratory
Kirtland AFB, New Mexico

M. Kristiansen
Texas Tech University
Lubbock, Texas

and

T. Martin
Sandia National Laboratories
Albuquerque, New Mexico

PLENUM PRESS • NEW YORK AND LONDON

Library of Congress Cataloging in Publication Data

Opening switches / edited by A. Guenther, M. Kristiansen, and T. Martin.
    p. cm.—(Advances in pulsed power technology; v. 1)
Includes bibliographical references and index.
ISBN 0-306-42664-1
    1. Electric switchgear. 2. Pulse generators. 3. Energy storage. I. Guenther, Arthur Henry, 1931–   . II. Kristiansen, M. (Magne), date. III. Martin, T. H. (Thomas H.) IV. Series.
TK2831.O64 1987
621.31′—dc19   87-16611
                                                                                          CIP

© 1987 Plenum Press, New York
A Division of Plenum Publishing Corporation
233 Spring Street, New York, N.Y. 10013

All rights reserved

No part of this book may be reproduced, stored in a retrieval system, or transmitted in any form or by any means, electronic, mechanical, photocopying, microfilming, recording, or otherwise, without written permission from the Publisher

Printed in the United States of America

FOREWORD

Pulsed power technology, in the simplest of terms, usually concerns the storage of electrical energy over relatively long times and then its rapid release over a comparatively short period. However, if we leave the definition at that, we miss a multitude of aspects that are important in the ultimate application of pulsed power. It is, in fact, the application of pulsed power technology to which this series of texts will be focused. Pulsed power in today's broader sense means "special power" as opposed to the traditional situation of high voltage impulse issues related to the utility industry.

Since the pulsed power field is primarily application driven it has principally engineering flavor. Today's applications span those from materials processing, such as metal forming by pulsed magnetic fields, to commercial applications, such as psychedelic strobe lights or radar modulators. Very high peak power applications occur in research for inertial confinement fusion and the Strategic Defense Initiative and other historical defense uses. In fact it is from this latter direction that pulsed power has realized explosive growth over the past half century. Early thrusts were in electrically powered systems that simulated the environment or effects of nuclear weapons detonation. More recently it is being utilized as prime power sources for directed energy weapons, such as lasers, microwaves, particle beam weapons, and even mass drivers (kinetic energy weapons). Consequently, much of the activity and growth in this field is spawned by government-sponsored research. This activity is of considerable benefit to the public and private sector through the commercial application of pulsed power technology in such diverse areas as medicine: by flash x-rays; commercial safety concerns: by lightning simulation; sewage treatment: by ozone production. Pulsed power is truly a multifaceted technology of diverse and surprising applications.

Pulsed power today is perhaps at once a flourishing field, yet one treated in perception as if it were of an arcane nature. Its exponential growth and development since World War II has taken place in the large without the benefit of text books, archival journals, or, for the greater portion of this period, recorded meetings or conferences devoted to pulsed power technology in an interdisciplinary manner. Only since the initiation of the IEEE Pulsed Power Conferences in 1976 have we been able to adequately delineate the requirements and focus pulsed power research from an engineering and physics standpoint. The physical, chemical and material sciences are now developing an appreciation and understanding of the processes involved in this field of pulsed power. Several interesting technology areas are involved, such as high voltage, high current, high power, repetitive discharges, accelerators, magnetic insulation, the effects of materials in adverse environments, instrumentation and diagnostics, as well as the always apparent problems related to switching, insulators, and breakdown. These events direct attention to the need for a strong and vigorous educational program. This need is not just a result of an increased awareness of pulsed power as an opportunity, but of the significance and potential of pulsed power to solve many of today's and the future's high technology problems.

The aforementioned explosive growth of this technology and the lack of recorded technical information has as well inhibited the development of pulsed power standards. We are now beginning to rectify this oversight. Our recent increase in understanding of fundamental pulsed power data has benefited greatly from the development of high quality diagnostic instrumentation techniques which have been perfected and applied to the field within the last 10-15 years. These techniques include lasers as applied in diagnosing plasmas or their use as light sources to visualize electric field distributions or even as voltage and current sensors using electro-optic effects. From the materials world the alphabet soup of surface analysis techniques, SEM, TEM, XPS, AES, LEED, SIMS, SNMS, ESCA, LIMA, etc., to name a few, have enhanced our understanding of not only surfaces but the interplay of materials and insulators either under the influence of electric fields or as a result of discharges. Importantly, we now have at hand rather elegant fast time resolution (approaching 10^{-15} second) diagnostics to employ in unraveling the processes limiting the performance of complex systems.

This series, therefore, is intended for physicists and engineers who have occasions to use pulsed power. They are not texts devoted to mathematical derivations but a thorough presentation of the present state-of-the-art. They will also point the reader to other sources when further in-depth study is necessary. We will address pulsed systems as distributed parameter systems in which electrical characteristics and propagation are considered in a continuous sense. Comparisons will be made, where appropriate, to lumped parametric analyses, such as in pulse forming networks.

FOREWORD

However, we will avoid a long list of inductance and capacitance formulae. Furthermore, we will assume that the readers of these texts are familiar with the basic material which address those aspects of the pulsed power field. One should be able to apply a general knowledge of dynamic processes in the transient world and have little difficulty in understanding the material covered in this series and its application and importance in a systems sense. We also assume that the reader has familiarity with normal circuit elements and principles as they apply to the pulsed power world. It will then be seen that our purpose is to draw attention to the important aspects of system design, performance, and operation.

The editors intend that this series consist ultimately of about a dozen volumes. The first two volumes will concentrate on those aspects of the field which usually are of the greatest concern as regards performance and reliability, to wit, switching technology. Thus, volumes I and II will cover opening and closing switch technology and we refer you to the foreword in each volume for an in-depth analysis of its coverage. Other topics to be covered in later volumes will include Energy Storage; capacitor, inductor, mechanical, chemical (batteries, explosives, etc.), Voltage Multipliers; i.e., Marx generators, LC lines, Blumlein generators, etc.; Pulse Forming; Electrical Diagnostics; Components; and a volume on Supporting Analytical Tools such as field plotting codes, magnetic insulation codes, etc.; and certainly one or more volumes on Pulsed Power Applications.

With this array of intended publications on Advances in Pulsed Power Technology we hope to lay the foundation for the field for many years to come. In addition, we will have the pleasure of publishing a volume containing the important pioneering and very tractable works of J.C. "Charlie" Martin and his co-workers of the Atomic Weapons Research Establishment, Aldermaston, Redding UK. We are sure this volume will be one coveted not only by those working in the field but also by students of the history of technology who will be able to use his contributions and style as a model for the rapid application of the scientific method to a new field with the insightfulness of an Edisonian empiricist. It promises to be enjoyable reading as well.

Certainly as this series of "Advances" are released new directions and aspects of the field will be revealed. It is expected by the series editors to dutifully address new "special power" issues such as space-based power, inductive storage systems employing repetitive opening switches, materials in adverse environments, etc., as they reach appropriate stages of development or interest. Actually, we anticipate that the need for pulsed power information will outrun our ability to publish the appropriate volumes. This phenomenon is indicative of a vital and practical technology.

It has been our extreme pleasure to have been part of the pulsed power scene to date, a rather close knit community. Our undertaking this effort to document a chronology of pulsed power advances has been our pleasure. We express our sincere appreciation to the individual volume editors both presently identified and for those to come. If the readers have suggestions for subjects to be included in this series or would like to contribute book chapters or serve as volumes editors, we would be most receptive to talking with them.

A.H. Guenther
M. Kristiansen

PREFACE

Originally it was thought by the editors that one volume on high-power switching would be an appropriate goal. To our consternation, this was a gross underestimate of the status of high power switch development and the considerable interest that we encountered as well as generated while outlining the book. We therefore requested the able assistance of Dr. K.H. Schoenbach of the Department of Electrical Engineering at Old Dominion University to address a major aspect of the intended initial switching volume. When Dr. Schoenbach completed his suggested outline, we now had information and the portent of much additional data, adequate to fill at least two volumes. As a result, the series editors subdivided the switching field into opening switches (this volume) and closing switches (the next volume). Dr. Schoenbach's insight and input were most valuable in unravelling this important subset of pulsed power technology, as switching is rightfully considered the topic of paramount interest in most pulsed power systems and many applications.

High power switching is key in controlling energy in a wide variety of electrical systems. Recently the peak electrical power output obtainable from pulsed power systems has increased to about 100 TW (over 100 times the total U.S. electrical power generating capability) in a single pulse. The specific application being ion beam fusion research for inertial confinement fusion. This large peak electrical power output has provided much stimulation for developing efficient power conditioning for smaller, more reliable, multi-pulse devices. These new devices utilize switching techniques that will be described in this and at least one additional volume.

This opening switch volume should be especially useful to the scientific community. The lack of good opening switches, particularly repetitive ones, has long inhibited the beneficial use of inductive energy stores, to be contrasted with the ready availability of simple closing switches which has aided the general utility of capacitive energy stores. Since inductive energy stores are considerably more compact than similar capacitive energy stores, the availability of single, efficient and reliable opening switches is of paramount importance to the power conditioning community. This first volume addresses that main problem area extensively. As such it will inform the novice and expert alike about the diversity of opening switch technology. At the same time the book furnishes detailed information for the more experienced reader. This goal is achieved since each chapter provides background information for the particular switching subject and then continues into more recent and important developments.

This volume discusses as well some opening switches that were abandoned, forgotten or reached an apparent limit of utility. As a result of new advances we believe that these switches may prove useful for some power conditioning applications and the reader is encouraged to review the supplied materials to ascertain if these concepts could meet their specific requirements, particularly their serial use in conjunction with other faster opening devices.

A brief overview of the book chapters should now be helpful in allowing the reader to choose those chapters of principal interest. This volume need not be read sequentially.

The first portion of Chapter One, "Inductive Energy Storage Circuits and Switches," by E. Honig, provides a basic and broad overview of present inductive store-opening switch combinations and is recommended for a first reading. Honig points out that opening switches are not necessarily less efficient than closing switches. However, the circuits in which the opening switches are used, inherently cause the opening switches to dissipate a larger portion of a circuit's total energy when compared to closing switches. Therefore, opening switches must always be considered in the context of the circuit in which they are employed. Chapter One then continues and discusses vacuum opening switches and semiconductor opening switches with an application example.

Chapter Two, "Diffuse Discharge Opening Switches," by K. Schoenbach and G. Schaefer, outlines a low pressure gas discharge concept that operates by changing the ionization source strength and allowing the gas recombination rate to "open" the switch. Electron beams and optical radiation can be used to control the diffuse discharges while special gas mixtures with specific attachment/detachment cross section can provide desirable closing, conduction and opening characteristics. Considerable information is now available for switch designers; however, the physics issues indicate that even better switches are likely if better gas mixtures were known. In this regard one employs "dielectric engineering."

PREFACE

Chapter Three, "Low-Pressure Plasma Opening Switches," by R. Schumacher and R. Harvey, discusses this interesting operating pressure region. The discussion leads from first principles into the magnetically switched Crossed Field Tube and the electrostatically controlled Crossatron Modulator switch. An example is given for a 100 kV, 10 kA interruptor tube. These devices interestingly have a repetitive operating capability.

As a transition between the fundamentals of inductive storage systems that may be repetitively operated usually through some low pressure opening switch and high current, high power/energy single pulse applications, one needs to more fully understand the recovery of dielectrics. To this end, in Chapter Four, "Recovery of Dielectric Strength in Gases", R. DeWitt presents in some detail a description of those processes which govern recovery, including space charge and many rate dependent effects which control and regulate electron-density changes.

Chapter Five, "The Plasma Opening Switch," by R. Commisso, G. Cooperstein, R. Meger, J. Neri, P. Ottinger, and B. Weber, describes a relatively new type of switch that has, since the chapter was initially written, opened several megamps in less than ten nanoseconds. This type of switch is somewhat difficult to use since it must be located close to the dissipative load and must as well operate in a vacuum. If these criteria are met, then this switch can provide current rise rates of 10^{15} amps/sec, employing a plasma erosion scheme.

Chapter Six, "The Reflex Switch: A Fast Opening Vacuum Switch," by L. Demeter, describes a multi-element high-current vacuum diode that can change impedance dramatically to control hundreds of kiloamps at megavolt potentials.

Chapter Seven, "Plasmadynamic Opening Switch Techniques," by P. Turchi, provides a general theoretical background for opening switches. Starting with Maxwell's equations, he derives general basic relationships for opening switches.

Chapter Eight, "Fuse Opening Switches for Pulse Power Applications," by R. Reinovsky, discusses the utility of metallic fuses as opening switches. He starts with basic circuit theory and then develops it into a description of an operating system that opens currents exceeding 25 megamps in a few microseconds, useful in transferring capacitively stored energy into low inductance stores as a power amplifier.

Chapter Nine, "Explosively-Driven Opening Switches," by B. Turman and T. Tucker, describe fuse type opening switches that are driven by high explosives instead of relying on the electrical circuit to drive a conductor to high velocities. Practical limits to the system size are shown to exist.

Chapter Ten, "Dielectric Breakdown in Solids," by A. Jonscher introduces the reader to solid breakdown phenomena. This chapter is useful on its own for describing the breakdown processes in solids. It is included in this book to provide the reader with information about the important breakdown phenomena that occur in solid state opening switches. The chapter also serves to start a transition to the next volume on closing switches.

Chapter Eleven, "Solid State Opening Switches," by M. Kahn, discusses the application of nonlinear solid state materials as opening switches. The mobility of the conducting carriers are decreased as the semiconductor is heated. Simple repetitively switched devices can be provided by using these principles, albeit at lower power and energy levels.

Chapter Twelve, "Mechanical Switches," by W. Parsons, reviews past and present developments of mechanically-operated switches. Surprisingly, mechanical opening switches have operated at 750 kV and currents in excess of 1 MA. This type of switching is at a relatively high level of development when compared to other opening switch types. Various principles of breaking circuits along with commercially available breaker units are discussed.

This book is oriented toward the reader who desires to understand the principles of different types of opening switch operation and gain the ability to assess the present limitations on ultimate performance. The editors hope you agree with them that there was enough useful information available on opening switches to suggest that an opening switch volume be published. Our intention is that this book will provide the basics for a wider use of opening switches and inductive energy stores as new power conditioning circuits.

The next volume will primarily address gas closing switches. When compared to opening switches, the closing switch area suffers from too much information. This next volume must be condensed, highlighted, and summarized because of the extremely large amount of research and the great number of reports in this relatively well-studied area.

Our most sincere appreciation goes to Marie Byrd who was responsible for the typing of this volume and who suffered with us during this learning period.

A. Guenther
T. Martin
M. Kristiansen

CONTENTS

Inductive Energy Storage Circuits and Switches   1
    E.M. Honig
    (Los Alamos National Laboratory)

Diffuse Discharge Opening Switches 49
    K.H. Schoenbach
    (Old Dominion University)
    G. Schaefer
    (Polytechnic University, NY)

Low Pressure Plasma Opening Switches 93
    R.W. Schumacher and R.J. Harvey
    (Hughes Research Laboratories)

Recovery of Dielectric Strength of Gases 131
    R.N. DeWitt
    (Naval Surface Weapons Center/Dahlgren)

The Plasma Erosion Opening Switch 149
    R.J. Comisso, G. Cooperstein, R.A. Meger,
    J.M. Neri, P.F. Ottinger
    (Naval Research Laboratory)
    B.V. Weber
    (JAYCOR)

The Reflex Switch: A Fast-Opening Vacuum Switch 177
    L.J. Demeter
    (Physics International Company)

Plasmadynamic Opening Switch Techniques 191
    P.J. Turchi
    (RDA Washington Research Laboratory)

Fuse Opening Switches for Pulse Power Applications 209
    R.E. Reinovsky
    (Air Force Weapons Laboratory)

Explosively-Driven Opening Switches 233
    B.N. Turman and T.J. Tucker
    (Sandia National Laboratory)

Dielectric Breakdown in Solids 257
    A.K. Jonscher
    (Royal Holloway and Bedford
    New College, England)

Solid State Opening Switches 273
    M. Kahn
    (Naval Research Laboratory)

Mechanical Switches 287
    W.M. Parsons
    (Los Alamos National Laboratory)

List of Contributors. 307

Index 309

INDUCTIVE ENERGY STORAGE CIRCUITS AND SWITCHES*

Emanuel M. Honig

Los Alamos National Laboratory

Los Alamos, New Mexico 87545

INTRODUCTION

   The purpose of an opening switch is simply to stop the flow of current in the circuit branch containing the switch. Prior to this action, of course, the opening switch must first conduct the current as required--that is, operate as a closing switch. To accomplish current interruption, the opening switch must force the current to transfer from the switch to a parallel circuit branch (e.g. a load) and then withstand the voltage generated by the current flowing through the load. If there is nothing in parallel with the switch branch, then the opening switch can interrupt the current only by absorbing all of the energy stored in the circuit inductance and recovering against the open circuit voltage of the current source. The severity of the switching problem depends upon the makeup of the rest of the circuit. Therefore, opening switches must be considered in the context of the circuit in which they are used.

---

*   The submitted manuscript has been authored by an employee of the University of California, operator of the Los Alamos National Laboratory under Contract No. W-7405, ENG-36 with the US Department of Energy. Accordingly, the US Government retains an irrevocable, nonexclusive, royalty-free license to publish, translate, reproduce, use or dispose of the published form of the work and to authorize others to do the same for US Government purposes.

Opening switches are used to provide fault current protection, to sharpen the current pulse of a capacitive discharge, or to enable the transfer of energy from an inductive energy store to a load. Examples of the first application include HVDC transmission lines and circuits with high impedance devices such as magnetrons and traveling wave tubes which normally operate with high voltages impressed across them and conduct only a small dc current. An opening switch in series with such devices, Fig. 1, can prevent the flow of large fault currents should a short circuit develop within the device.[1,2] To interrupt the fault current, the opening switch must absorb all of the energy stored in the circuit's stray inductance and then withstand the high voltage remaining on the source as shown in Fig. 2.

The more difficult applications for opening switches are those which require the switch to carry a large current when closed and then on command to transfer the current to a parallel load. In effect, the opening switch inserts the load into the circuit so that energy can be transferred from the source to the load. These applications include the pulse sharpening technique for a capacitive discharge as well as the standard inductive energy storage system.

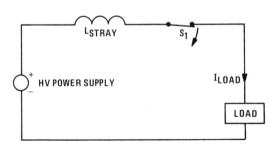

Fig. 1. Opening switch used for fault-current protection.

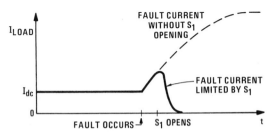

Fig. 2. Current waveform showing effectiveness of opening switch in limiting fault current magnitude.

In the pulse sharpening application, Fig. 3, a load requiring a fast-rising current pulse is bypassed with a closed opening switch, S1, during the initial part of a capacitive discharge. Near peak current, the switch opens to force a rapid transfer of the current to the load, as shown in Fig. 4. This scheme is really a form of inductive energy storage, although the inductor may not be apparent as it may only be the stray inductance of the connections between the capacitor and the opening switch.[3] Recently, this technique was used with a 9.5 MJ capacitor bank at the Air Force Weapons Laboratory to transfer up to 14 MA to a resistive load.[4]

The standard inductive energy storage system, Fig. 5, is used to supply power in the form of a large single pulse or a train of high power pulses. Energy is transferred from the inductive store to the load each time the opening switch operates, Fig. 6. Inductive energy storage systems are discussed in considerable detail in Ref. (1).

The need for such systems is now emerging from major programs within the DOE and DoD communities.[5-8] They include repetitively-pulsed particle accelerators and lasers for industry, nuclear fusion, and defense; impulse radiation sources for nuclear-weapons-effects simulation; intense microwave sources for weapons and

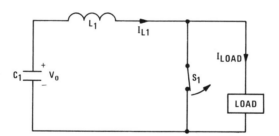

Fig. 3.   Opening switch used for pulse-sharpening of the current from a capacitive discharge.

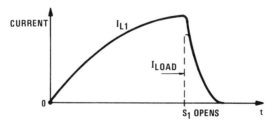

Fig. 4.   Waveforms of capacitive discharge current and load current.

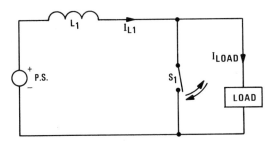

Fig. 5. Opening switch used in an inductive energy storage system to transfer energy to a load.

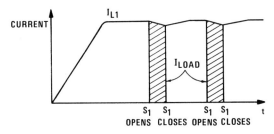

Fig. 6. Simplified waveforms of the storage coil current and load current for an inductive energy storage system.

weapons-effects simulation; high power radar; and induction heating systems. The importance of the many applications and the lack of a suitable technology to meet these needs led the U.S. Army Research Office to sponsor a series of workshops addressing these problems at Tamarron, Colorado.[8-11] Table 1 gives a comparison of the projected needs of several representative applications and is adapted primarily from the proceedings of the 1981 Repetitive Opening Switch Workshop at Tamarron.[8] Pulse requirements for the various applications cover a wide range: kiloampere-megampere currents, kilovolt-megavolt voltages, nanosecond-millisecond pulse widths, and pulse repetition rates of 1-to-30,000 pulses per second (pps).

Many of the systems under consideration must operate repetitively and have a low output duty factor (ratio of pulse width to pulse interval). The average power required in a pulse train will be an equally low fraction of the peak power in the pulse, Fig. 7. Rather than use a power supply rated at the peak pulse power, P_{PEAK}, to drive the load directly, as shown in Fig. 8(A), it is generally more practical -- perhaps even mandatory -- to use a much smaller power supply rated at the average pulse train power, P_{AVG}, and an energy storage system capable of storing the energy of at least one output pulse, as shown in Fig. 8(B). The energy store is charged by the smaller power supply, rated at P_{AVG}, during the interval between pulses and then discharged to give the required

Table 1

TYPICAL APPLICATIONS[8]

Application	Voltage (MV)	Current (MA)	Rep. Rate (kpps)	Load Type (*)	Pulse Energy (MJ)
Advanced Test Accelerator	0.1-1	0.01-0.1	10	C	0.002-0.5
Modified Betatron	0.5	0.2	0.001-10	L	100
Lasers	0.01-0.1	0.001-0.03	0.01-0.1	R or C	0.001-0.04
EM Guns	0.002-0.02	0.1-5	0.001-0.05	R and L	5-500
Inertial Fusion	3	0.1	0.01	C	0.1-3
EMP Simulator	1-5	0.01	0.1	C	0.1-0.5

* C = Capacitive; R = Resistive; L = Inductive

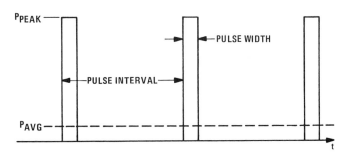

Fig. 7. Typical repetitive pulse train illustrating the difference between peak and average power.[1]

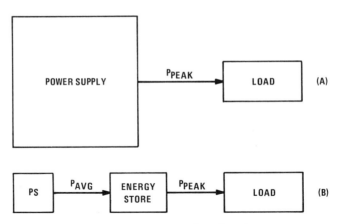

Fig. 8. Simplified block diagrams of pulsed-power systems comparing a power supply rated at the peak pulse power (A) and a power supply rated at the average power in a pulse train (B) to illustrate the need for energy storage.[1]

peak power, P_{PEAK}, to the output pulse. An alternative mode of operation is to charge the energy store with the total energy of a train of pulses and then recharge it between pulse trains. This requires a larger energy store but may allow for a much smaller power supply.

The advantages of using an energy store increase as the repetition rate decreases. In the limit, single shot operation, an energy store is almost mandatory. Therefore, energy storage and transfer systems are important, not only for their potential savings in overall system size, weight, and cost, but also because the enormous peak power requirements of many applications make power supplies rated at the peak pulse power completely impractical.

The energy storage systems generally used are capacitive, inductive, chemical (batteries and high explosives), and inertial (rotating machines, possibly augmented with flywheels). The salient features in comparing these systems are energy storage density, storage losses, charge and discharge rates (which in turn determine the minimum energy transfer times and peak power capabilities), and appropriate scaling laws (cost, size, etc.). Table 2 and Fig. 9 show a comparison of these storage methods, based primarily on papers by Nasar[5] and by Rose.[6] High explosives have the highest energy density and "short" energy-release times, but they are limited to single-shot operation and require auxiliary equipment to convert the chemical energy to electrical energy. Batteries have a high energy storage density but a low power capability, requiring long charge and discharge times. Inertial storage has a high storage density and a moderate power capability. Capacitors

INDUCTIVE ENERGY STORAGE CIRCUITS AND SWITCHES

TABLE 2

COMPARISON OF ENERGY STORAGE METHODS[1]

Storage Device	Energy Density (MJ/m^3)	Energy/Weight (J/kg)	Typical Transfer Time
Capacitors	0.01-1	300-500	microseconds
Inductors	3-40	10^2-10^3	microsecond-milliseconds
Batteries	2000	10^6	minutes
Explosives	6000	5×10^6	microseconds
Flywheels	400	10^4-10^5	seconds

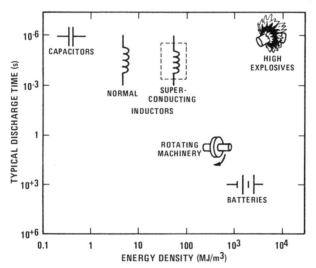

Fig. 9. Comparison of various energy storage methods.[1]

have the highest electrical discharge capability but a relatively low energy storage density. Only inductive storage has both a high energy density and a high electrical power capability. Inductive storage also has a decreasing ratio of cost per unit energy as size increases[12] (due to the effect of mutual inductance as new turns are added) and can be made essentially lossless with superconducting materials.[13,14] Because non-superconducting or normal coils are a "lossy" storage medium, energy must be stored in such coils for times less than about one L/R time constant (where L is the coil inductance and R is its resistance) or else the coil itself will dissipate too much energy to be practical.[15] The need to charge a normal coil quickly requires that the power supply be much larger than for other storage systems and makes the prime power unit about equal in significance to the coil in such systems. Nevertheless, normal coil systems are still receiving considerable attention - partly because of the ease with which the coil can be charged from a low-loss, high-energy-density inertial store and partly because of the attractive compromise offered by liquid-nitrogen-cooled aluminum coils.

For background information in the area of high power opening switches, a number of texts and review articles are available. General pulsed power techniques and applications are discussed by Rose.[6] Industrial switches and circuit breakers are discussed in the textbooks by Lee[16] and Flurscheim.[17] High power switches are the topic of the Tamarron workshops[8-11] and are reviewed by Burkes et al.[18] Opening switches for inductive storage are reviewed by Honig[1] and by Schoenbach et al.[19]

In summary, significant advances in many pulsed power systems can be made by the further development of opening switches. The wide range of operational requirements and applications require that the research and development of such switches be carried out on a broad front.

OPENING SWITCH AND TRANSFER CIRCUIT FUNDAMENTALS[1]

Basic Types of Opening Switches

Opening switches can be conveniently divided into two basic classes, direct-interruption and current-zero, depending upon whether the voltage which forces the current to transfer out of the switch is developed internal or external to the switch. Which type of switch is better to use in a particular application is often determined by the system characteristics, such as peak current, peak voltage, conduction time, and resistance and inductance of the load.

Direct-Interruption Opening Switches. Switches that can develop the transfer voltage internally, through a mechanism in which the generalized switch impedance is increased, are called direct-interruption or "true" opening switches. Examples include dc circuit breakers, fuses, crossed-field devices, dense plasma focus devices, superconducting switches, and laser sustained and electron beam sustained diffuse discharge switches. Many of these types of switches are discussed in other chapters of this book.

Figure 10 shows a circuit diagram of a basic inductive energy storage and transfer system with a direct-interruption opening switch. Simplified waveforms typical of the transfer process are shown in Fig. 11. The current, I_0, in the energy storage coil, L_1, initially flows through the closed opening switch, represented by its time-varying impedance $R(t)$. To transfer energy to the load, R_L and L_2, the switch impedance, $R(t)$, is raised rapidly at time t_1 to a value R_{SWITCH} much greater than the load impedance to "choke off" the current flow and force the switch current (loop current I_1) to transfer to the load (loop current I_2). The final current, I_2, is less than the initial current, I_1, because energy is dissipated in the opening switch during the switching process. The switch must conduct the coil current for the full charge and storage times, undergo a large impedance change, and then withstand the high recovery voltage immediately afterward. The load voltage risetime depends directly on the switch impedance risetime. After switch opening, the load pulse decays exponentially with a time constant of $(L_1 + L_2)/R_L$ as energy is transferred from the storage inductor to the load. To terminate the load pulse at time t_2, the opening switch must reclose to short-circuit the coil and return the system to the storage mode. If load pulses are required repetitively, this cycle must be repeated for each pulse. The circuit shown in Fig. 10 is called a resistive transfer circuit because the high impedance of the opening switch (or an optional parallel resistor, not shown), is the element forcing the current to transfer to the load.

Current-Zero Opening Switches. A second method to achieve current interruption is to use a voltage source external to the switch to drive the switch current to zero and allow the switch to open. Switches that recover rapidly under such conditions are called current-zero opening switches. Their main application is in ac systems, where current zeros occur naturally twice each cycle,

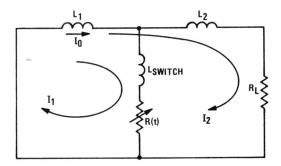

Fig. 10. Schematic circuit diagram[1] of a basic inductive energy storage and transfer system with a direct-interruption opening switch of impedance $R(t)$.

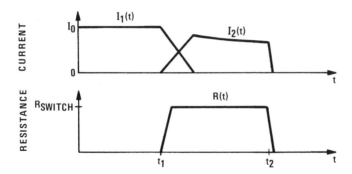

Fig. 11. Simplified waveforms[1] of opening switch current, $I_1(t)$, load current, $I_2(t)$, and switch resistance, $R(t)$, for a resistive transfer circuit (Fig. 10) with a direct-interruption opening switch.

but they may also be used in dc systems if an auxiliary circuit is used to produce a momentary current zero. Examples of current-zero switches include solid-state thyristors, vacuum switches, hydrogen thyratrons, and the liquid metal plasma valve. Thyristors and vacuum switches are discussed later in this chapter.

In dc systems, the current zero in the opening switch must be produced artificially, generally by discharging a capacitor through the opening switch in the direction opposing the main current. Fig. 12 shows a circuit diagram of a general inductive energy storage system with a current-zero opening switch. Waveforms showing the details of switch opening are shown in Fig. 13. Initially, the storage coil current, I_0, flows through the current-zero opening switch, S_1. To create a current zero in S_1, switch S_2 is closed to initiate the flow of counterpulse current, I_{CP}. When the opening

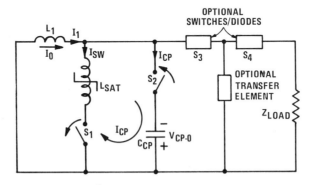

Fig. 12. Circuit diagram[1] of generalized inductive storage system with a counterpulsed current-zero opening switch, S_1.

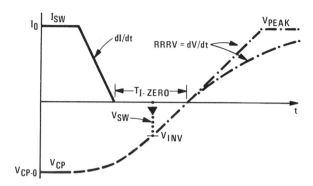

Fig. 13. Simplified waveforms[1] of the opening switch current, the counterpulse capacitor voltage, and the switch recovery voltage for the circuit of Fig. 12.

switch current, I_{SW}, reaches zero, the series saturable reactor, L_{SAT}, comes out of saturation and introduces a large increase in inductance to the opening switch branch, preventing any further increase in current, I_{CP}. The switch current, I_{SW}, is effectively "frozen" at zero current for a time T_{I-ZERO}. The duration of the current-zero time, T_{I-ZERO}, is determined by the volt-second rating of the saturable reactor and the average voltage on C_{CP} during time T_{I-ZERO}. Of course, switches with reverse blocking capability, such as thyristors, may not need a series saturable reactor. For the opening switch to open electrically, the current-zero time provided by the counterpulse technique must be greater than the switch recovery time. When the switch recovers, sometime during T_{I-ZERO}, it first sees an inverse recovery voltage, V_{INV}, equal to the voltage, V_{CP}, remaining on C_{CP} at that instant. The switch recovery voltage then tracks the capacitor voltage as it decreases rapidly towards zero for the remainder of T_{I-ZERO}. After the recovery voltage and capacitor voltage reverse polarity, the rate of rise of recovery voltage, RRRV (or dV/dt), and the peak recovery voltage, V_{PEAK}, are determined by the type of transfer circuit used and the type of load being driven. The two general patterns for the recovery voltage waveform following polarity reversal are linear or exponential, as shown in Fig. 13. Therefore, for a current-zero switch, the important operational parameters are the initial current level, I_0; the length of conduction time; the rate of current reduction, dI/dt, before current zero; the length of current-zero time, T_{I-ZERO}; the initial inverse recovery voltage, V_{INV}; the rate of rise of the recovery voltage, RRRV, in the forward direction; and the peak recovery voltage, V_{PEAK}.

Transfer Circuits

Any discussion on opening switches would be incomplete without describing the circuits in which they are used. In applications where the opening switch initiates the transfer of energy from an inductive store to a load, the circuits can be classified according to the type of circuit element causing the transfer action, with resistive and capacitive elements being the most common.[20]

Resistive Transfer Circuit. The resistive transfer circuit, Fig. 10, has already been discussed briefly. The transfer resistance may be the resistance R(t) of the direct-interruption opening switch by itself or the resistance of the parallel combination of the opening switch and a transfer resistor. Whether a direct-interruption opening switch requires the aid of a transfer resistor is generally determined by the peak system voltages and total switch dissipation. The efficiency of the resistive transfer circuit depends upon the load--whether it is predominantly resistive or inductive. To understand the operation of the resistive transfer circuit for various loads, it is useful to first consider its operation for the two special cases of a purely resistive load and a purely inductive load.

Figure 14 shows an ideal resistive transfer circuit having zero coil resistance, zero switch ON-state resistance, and no stray inductances in series with the switch or load. The assumed switch $R(t)$ is a step function where the switch impedance changes from 0 to R_{SWITCH} at $t = 0$.

For the ideal resistive transfer circuit (Fig. 14) with a purely resistive load, R_{LOAD}, the amount of current, I_{LOAD}, transferred to the load depends upon the ratio of R_{LOAD} to the OFF-state resistance, R_{SWITCH}, of the opening switch (Fig. 11) and is simply given by the current-divider equation,

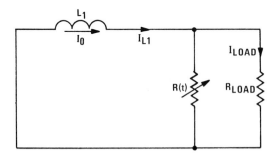

Fig. 14. Ideal resistive transfer circuit with purely resistive load.[1]

$$I_{LOAD} = I_{L1}(t)[R_{SWITCH}/(R_{SWITCH} + R_{LOAD})], \quad (1)$$

where $I_{L1}(t)$ is the current flowing in the storage coil, beginning with an initial value of I_0.

For inductive storage systems, a good measure of the system operation is to compare the energy in the load pulse(s) to the decrease in stored energy during the transfer cycle(s). Since the difference between the transferred energy and the decrease in stored energy can be attributed to energy loss mechanisms, this comparison gives a good indication of circuit efficiency. If the energy transferred to the load element is defined as E_{LOAD}, then a measure of the overall circuit efficiency can be found by defining the operational efficiency, η, to be

$$\eta = E_{LOAD}/\Delta E_{L1}, \quad (2)$$

where ΔE_{L1} is the decrease in the stored energy in the storage coil L_1.

Therefore, for the resistive transfer circuit with a purely resistive load, the ideal operational efficiency, $\eta_R(ideal)$, is given by

$$\eta_R(ideal) = R_{SWITCH}/(R_{SWITCH} + R_{LOAD}), \quad (3)$$

while the switch dissipation, E_{SWITCH}, for the ideal case is

$$E_{SWITCH}(ideal) = E_{LOAD}R_{LOAD}/R_{SWITCH}. \quad (4)$$

These equations show that most of the current can be transferred to the load at a high operational efficiency if R_{SWITCH} is much greater than R_{LOAD}. To achieve good efficiencies in actual practice, resistive transfer circuits must also minimize the coil resistance and the ON-state switch resistance, as detailed in Ref. 1. In addition, the stray inductances of the switch and of the load must also be minimized as these force the opening switch to dissipate even more energy during the transfer.

An inductive load will resist the efforts of the opening switch to transfer current to the load, causing considerable energy to be deposited in the switch. For an ideal resistive transfer circuit with a lossless storage coil, L_1, an ideal opening switch, S_1, a transfer resistor, R, and a purely inductive load, L_2, as shown in Fig. 15, the time-dependent equations for the storage coil current, $I_{L1}(t)$, and the load coil current, $I_{L2}(t)$, according to Carruthers,[20] are given by

$$I_{L1}(t) = I_0[1/(1 + X)][1 + X\exp(-t/\tau)], \quad (5)$$

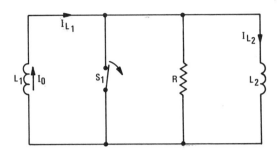

Fig. 15. Resistive transfer circuit with a purely inductive load.

$$I_{L2}(t) = I_0[1/(1 + X)][1 - \exp(-t/\tau)], \tag{6}$$

where the load to storage inductance ratio, X, is given by

$$X = L_2/L_1, \tag{7}$$

the exponential time constant, τ, is

$$\tau = L^*/R, \tag{8}$$

and the effective inductance, L^*, of L_1 and L_2 in parallel is

$$L^* = L_1 L_2/(L_1 + L_2). \tag{9}$$

The transfer process can also be analyzed using the principle of conservation of flux to determine the various current and energy ratios[1].

The maximum current, I_2, transferred to the load is

$$I_2 = I_0/(1 + X), \tag{10}$$

where I_0 is the initial current flowing in the storage coil and X is defined by Eq. (7).

The energy, E_2, transferred to the load is

$$E_2 = E_0 X/(1 + X)^2, \tag{11}$$

where E_0 is the initial energy stored in L_1.

The energy, E_R, dissipated in the resistive transfer element is given by

$$E_R = E_0 X/(1 + X) \tag{12}$$

and the ratio of transferred to dissipated energy is

$$E_R/E_2 = (1 + X). \tag{13}$$

Appropriate ratios are plotted as a function of X in Fig. 16. The magnitude of the transfer resistance has no effect on these ratios; it merely determines the peak voltages generated during the transfer and the speed at which the transfer can occur. Under conditions of maximum energy transfer, where $L_2 = L_1$ (or $X = 1$), it can be seen that only 25 per cent of the initial stored energy is transferred to the load, Eq. (11), and that 50 per cent is dissipated in the transfer resistor Eq. (12). It is apparent from Eq. (13) that the transfer element dissipates about as much energy as is transferred when X is small and dissipates twice as much as the transferred energy when the inductors are matched.

The overall or operational efficiency can be obtained by comparing the energy transferred to the load (E_2) with the energy transferred from the storage coil (E_2+E_R). For the ideal resistive transfer circuit with a purely inductive load and a lossless storage coil, the operational efficiency, η_L, as defined by Eq. (11), is given by

$$\eta_L = E_2/(E_2+E_R) = 1/(2+X). \tag{14}$$

Therefore, the operational efficiency can never exceed 50 per cent, even for very small values of the inductance ratio X, because the switch or transfer resistance always dissipates at least as much energy as is transferred. As X increases, η_L decreases even further.

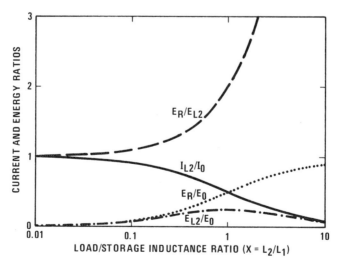

Fig. 16. Current and energy ratios for operation of the resistive transfer circuit with a purely inductive load.

The fundamental energy dissipation in the transfer resistor imposed by Eq. (13) is a significant disadvantage for this method of energy transfer. If the transfer is caused solely by the impedance increase of the opening switch, then this energy must all be dissipated within the switch itself. This is a severe penalty to impose on a switch trying to interrupt the current and simultaneously withstand the recovery voltage.

The penalty is even worse when the inductance of the switch is included. The energy, $E_{L-SWITCH}$, stored in the stray inductance, L_{SWITCH}, of the opening switch circuit branch must also be dissipated in the switch because it is the only dissipative element in the circuit. Therefore,[1,21,22] the energy dissipation, E_{SWITCH}, in a direct-interruption opening switch transferring current to a purely inductive load, L_2, is

$$E_{SWITCH} = E_{L-SWITCH} + (1 + X)E_{L2}, \qquad (15)$$

where E_{L2} is the energy transferred to L_2.

This is the minimum amount of energy that must be dissipated in the opening switch. If the load also has a resistive component, then the opening switch has to work that much harder to cause the current to transfer and increased switch dissipation occurs. For such loads, the analysis of switch dissipation and operational efficiency is best handled with a circuit analysis code on a computer (e.g. NET-2 or Spice).

The implications of these fundamental relations for switches in inductive energy storage systems are clear. In resistive transfer circuits, the energy associated with the stray inductances must be fully considered and the energy dissipation required by Eq. (15) dealt with. The switch and load inductances may have enough effect upon the operation of the opening switch and the circuit transfer efficiency to make some direct-interruption opening switch schemes impractical. On the other hand, current-zero switches can be coupled with the high transfer efficiency possible with counterpulse techniques and capacitive transfer circuits to provide an approach that may be more practical, as will be discussed in the next section.

<u>Capacitive Transfer Circuit</u>. As the name implies, capacitive transfer circuits have a capacitor as the circuit element which forces the current to transfer to the load. When the opening switch opens, the current is first transferred to the capacitor. The rising voltage on the capacitor then causes the current to transfer to the load.

Either a direct-interruption switch or a current-zero switch may be used with the transfer capacitor. Usually, current-zero

INDUCTIVE ENERGY STORAGE CIRCUITS AND SWITCHES 17

switches are used and the transfer capacitor also serves as the counterpulse capacitor. In this case, it is very important to distinguish between the energy needs of a counterpulse capacitor for opening switch duty and the energy storage needs of a transfer capacitor in driving a particular type of load. For example, to transfer energy from an inductive store to an equally-sized inductive load, the transfer capacitor must be sized to handle one-half of the energy initially held in the storage inductor. In contrast, the capacitive energy needed to produce an adequate counterpulse in a current-zero opening switch is usually only a small fraction of this amount. In fact, confusion about these two different capacitor requirements is probably the main reason that counterpulsed current-zero switches operating in a capacitive transfer circuit were overlooked for so long as possible candidates for high-repetition-rate opening switches.[1]

With a resistive load, the capacitive transfer circuit has two possible modes of operation, depending upon whether a load isolation switch is used. If the transfer capacitor, C, and load resistor, R, are directly in parallel (Fig. 17), a relatively-constant current is supplied by the storage coil to the RC load. However, the current pulse in R has a relatively-long risetime of about 3RC, as shown in Fig. 18. This circuit is usually not attractive because of the long risetime.

Much faster load pulse risetimes can be obtained by isolating the resistive load from the transfer capacitor with a load isolation switch, such as S_2 in Fig. 19, and charging the capacitor to a voltage about equal to the expected load voltage ($\sim I_0 R_{LOAD}$) before closing S_2 to connect the load to the circuit and initiate the load pulse. The load isolation switch also insures that all of the counterpulse current produced by the capacitor flows through the opening switch. The load current then rises quickly (Fig. 20),

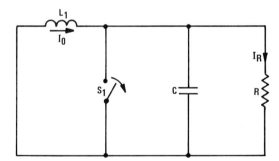

Fig. 17. Capacitive transfer circuit with the resistive load directly in parallel with the transfer capacitor.

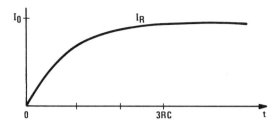

Fig. 18. First part of load current pulse obtained with the circuit of Fig. 17.

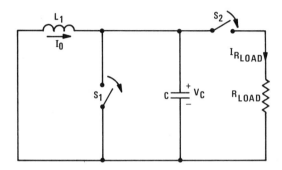

Fig. 19. Capacitive transfer circuit with the resistive load R_{LOAD} isolated from the transfer capacitor C with a load isolation switch, S_2.

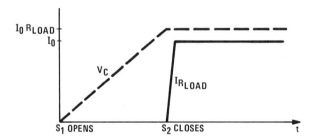

Fig. 20. Waveforms of transfer capacitor voltage and load current for the circuit of Fig. 2-10.

with the risetime limited primarily by the stray inductance in series with the load. The actual dependency of the load pulse risetime as a function of various circuit parameters is discussed in detail in Ref. 1. When the load current reaches the same level as the storage coil current, the current in the capacitor branch is zero. At this instant, the capacitor can be disconnected from the circuit by using a current-zero switch in series with the capacitor. This switch may also serve as the counterpulse start switch.

If the load is inductive (Fig. 21), the transfer capacitor merely sets up a resonant circuit condition between the storage inductor, L_1, and the load inductor, L_2. The current transfers smoothly into L_2, with the capacitive voltage V_C reaching its peak midway during the transfer (Fig. 22). The current waveforms shown in Fig. 22 depict the matched inductance case, with $L_1 = L_2$. In this case, all the energy stored in L_1 is transferred to L_2 during one transfer cycle. The energy is then trapped in L_2 by reclosing S_1. During the resonant transfer, the transfer capacitor must handle a peak energy of $(1 + L_2/L_1)/4$ times the energy transferred to the load, or a total of one-half the initial energy when $L_1 = L_2$. This is generally much larger than required just for counterpulsing, leading to a common misunderstanding of this technique in general.

The differences in performance for the resistive and capacitive transfer circuits are clearly evident in the case of a purely inductive load. As discussed earlier, the resistive transfer circuit operates very inefficiently with this type of load. The capacitive transfer circuit, however, can operate quite efficiently with an inductive load and have the added bonus of reducing the demands on the opening switch.

A complete analysis of the capacitive transfer circuit with an inductive load (Fig. 21), is presented in Ref. 1. A similar analysis was first presented by Carruthers.[20] Letting $X = L_2/L_1$, the load to storage inductance ratio, the current, $I_1(t)$, in the storage coil, L_1, is given by

$$I_1(t) = I_0(1 + X \cos \omega t)/(1 + X), \tag{16}$$

the current, $I_2(t)$, in the load inductor, L_2, is

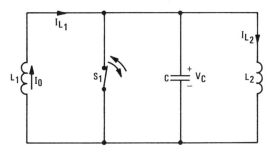

Fig. 21. Capacitive transfer circuit with a purely inductive load.

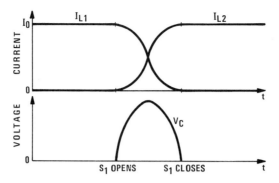

Fig. 22. Waveforms of storage coil current, load coil current, and transfer capacitor voltage for circuit of Fig. 21.

$$I_2(t) = I_0(1 - \cos \omega t)/(1 + X), \tag{17}$$

and the voltage, V_C, across the transfer capacitor, C, is

$$V_C(t) = L_2 I_0 (\omega \sin \omega t)(1 + X). \tag{18}$$

In these equations, the characteristic frequency, ω, of the system, is

$$\omega = (1/L^*C)^{1/2}, \tag{19}$$

where L^*, the equivalent inductance of L_1 and L_2 in parallel, is

$$L^* = L_1 L_2/(L_1 + L_2). \tag{20}$$

From these time-dependent equations, the current and energy ratios can easily be calculated. The ratio of the peak current, I_2, in L_2 to the initial current, I_0, in L_1 is the current transfer ratio

$$I_2/I_0 = 2/(1 + X), \tag{21}$$

which is twice that for the resistive transfer circuit, Eq. (10).

The ratio of the energy, E_2, transferred to L_2 to the initial energy, E_0, stored in L_1 is the energy transfer ratio

$$E_2/E_0 = 4X/(1 + X)^2, \tag{22}$$

which is four times better than with a transfer resistor, Eq. (11). The maximum energy transfer occurs when $L_2 = L_1$ (when $X = 1$) and all of the stored energy is transferred to the load in a single pulse.

INDUCTIVE ENERGY STORAGE CIRCUITS AND SWITCHES

The ratio of the maximum energy, E_C, handled by the capacitor, C, to the initial energy, E_0, is the capacitive energy ratio

$$E_C/E_0 = X/(1 + X), \qquad (23)$$

which is the same as the resistive energy ratio, Eq. (12). The key difference is that the transfer resistor dissipates its energy during the transfer while the capacitor stores its energy temporarily and then transfers it to the load during the pulse.

Finally, the ratio of capacitive energy, E_C, to the transferred energy, E_2, is

$$E_C/E_2 = (1 + X)/4, \qquad (24)$$

which is only one-fourth of the equivalent ratio for the resistive case, Eq. (13). Under conditions of maximum energy transfer (when X = 1), the capacitor must handle one-half of the energy initially stored in L_1, Eq. (23). This greatly diminishes the advantages of inductive storage for single shot operation. However, for repetitive operation in which only a small part of the total stored energy is transferred per pulse, the inductance ratio X will be much less than one and the capacitor size becomes much more favorable, equal to only one-fourth of the energy E_2 in the load pulse. The addition of such a capacitor is a small price to pay to achieve highly efficient operation with an inductive load. The current and energy ratios discussed above, Eqs. (21-24), are shown plotted as functions of the inductance ratio X in Fig. 23 and may be compared with the resistive transfer circuit results shown in Fig. 16.

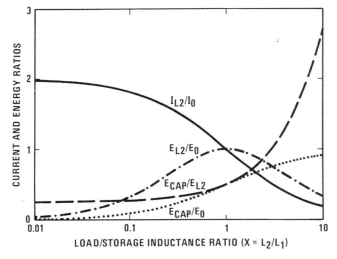

Fig. 23. Current and energy ratios for the capacitive transfer circuit with a purely inductive load.

Finally, the ratio of the residual energy, E_1, remaining in L_1 after the load pulse to the initial energy, E_0, stored in L_1 is

$$E_1/E_0 = (1 - X)^2/(1 + X)^2. \tag{25}$$

Adding the transferred energy, E_2, Eq. (22), to the residual energy, E_1, Eq. (25), gives the total energy, E_0,

$$\begin{aligned} E_1 + E_2 &= E_0[4X/(1 + X)^2 + (1 - X)^2/(1 + X)^2] \\ &= E_0[(1 + 2X + X^2)/(1 + X)^2] \tag{26} \\ &= E_0, \end{aligned}$$

demonstrating that no energy is lost during the transfer process and showing that the capacitive transfer circuit is 100 per cent efficient, in principle, with a purely inductive load. Of course, real capacitive transfer circuits will experience some energy loss in the switches (due to their forward conduction drop) and in the resistance of the conductors. However, these losses will be small compared to the energy dissipated in the opening switch and/or resistor, Eq. (15), when using a resistive transfer circuit with an inductive load.

In actual circuits with a resistive load, there will always be some inductance associated with the circuit connections. The use of a counterpulsed current-zero switch with such a load results in a capacitive transfer circuit with a predominantly resistive load. In this case, the output pulse risetime depends both on the inductance of the load and on the capacitor voltage just prior to the closing of the load switch. Operation of such a circuit and its analysis are discussed in complete detail in Ref. 1.

CURRENT-ZERO SWITCHES

Introduction

When fast opening switches are needed, one generally thinks of fuses or other direct-interruption switches with a fast opening mechanism. With such switches, the risetime of the load pulse directly follows the risetime of the switch impedance. In contrast, current-zero switches usually require a period of zero current (the current-zero time) of many microseconds to allow the switch to deionize and open. While current-zero switches do not at first appear to be fast opening switches, it is important to realize that there is no direct connection between the load pulse risetime and the current-zero time. The current-zero time is determined by the switch's deionization time, but the load pulse risetime depends only on the circuit parameters of the capacitor-

INDUCTIVE ENERGY STORAGE CIRCUITS AND SWITCHES

load discharge loop, as discussed later. Therefore, current-zero switches can be used to produce output pulses with risetimes that are many times less than the current-zero time required by the switch.

As the power requirements of fast pulse applications increase, one must consider the advantages offered by the use of current-zero switches. First, much of the technology for current-zero switches comes from the well-developed field of ac power, where power ratings and reliability are important. Solid-state thyristors and vacuum interrupter switches are good examples of developments in this field. Second, when compared to direct-interruption switches, the demands placed on current-zero switches are considerably reduced because the switch is allowed to open under conditions of zero current and relaxed recovery voltage stress. Finally, the counterpulse technique used with current-zero switches in a capacitive transfer circuit enables highly-efficient energy transfer to a load with a sizable inductive component without the energy dissipation penalties associated with direct-interruption switches, as discussed earlier in this chapter.

Natural applications for current-zero switches include inductive energy storage circuits with inductive loads and with repetitive-pulse resistive loads. In these applications, the counterpulse capacitor also serves as the transfer capacitor. The second application, delivering fast repetitive pulses to resistive loads, will be discussed in more detail later in this section.

Types of Switches

In principle, almost any type of switch that recovers when its current is reduced to zero can be used as a current-zero opening switch. However, because the size of the counterpulse capacitor is directly related to the current-zero time required, one is generally limited to those current-zero switches that have a fast recovery time. Current-zero switches for practical applications are mostly limited to vacuum interrupters, triggered vacuum gaps, and solid-state thyristors, as discussed below. Hydrogen thyratrons should also be considered for fast pulse applications. For very high-power non-repetitive applications for current-zero switches, ac breakers such as the air-blast, the SF_6-blast, and the SF_6-arc-spinner switches are suitable.

Vacuum Switches[1]. With their short deionization (recovery) time, demonstrated high-power interruption capability, and relatively low cost, vacuum switches are ideal candidates for many applications requiring current-zero switches. Mechanically-operated vacuum interrupters lend themselves well to single-pulse applications, while triggered vacuum gaps must be used if the pulse repetition rate exceeds about 20 pulses per second (pps).

A typical vacuum interrupter switch used in ac electric utility applications consists of two planar or disc electrodes, one fixed and the other movable, inside a vacuum envelope with a nominal gas pressure of 10^{-6} torr or less, as shown in Fig. 24.[23] The movable electrode is attached to an actuator mechanism through a set of bellows to maintain the vacuum. During normal operation in ac systems or during the charging cycle of an inductor, the contacts are mechanically closed to provide a low-resistance (10-50 $\mu\Omega$), metal-to-metal contact. Before the switch can be opened electrically, the mechanical actuator must first separate the electrodes. An arc is then drawn between the electrodes, sustained by the evaporation of metal vapor from the electrodes themselves. As the current passes through zero, naturally in ac systems or by a current counterpulse in dc systems, the power input to the arc decreases dramatically, allowing the plasma to recombine and the switch to recover. After current zero, vacuum switches can withstand a rate of rise of recovery voltage (dv/dt) as high as 24 kV/μs, faster than for any other type of high-power switch.[16]

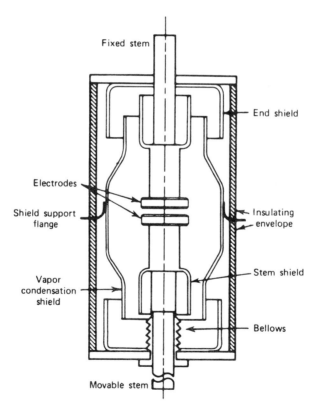

Fig. 24. Cross section of a typical vacuum interrupter.[23]

INDUCTIVE ENERGY STORAGE CIRCUITS AND SWITCHES

At the Los Alamos National Laboratory, vacuum interrupters have been the workhorse opening switch for inductive energy storage systems since the early 1970s. By 1975, current interruptions in excess of 30 kA with recovery voltages over 60 kV were demonstrated with a single vacuum interrupter.[24] Further work at Los Alamos investigated the parallel operation of vacuum interrupters,[25] water cooling of the electrodes,[26,27] and the counterpulse technique.[28] In a series of proof tests[29] of switching systems for the Tokamak Fusion Test Reactor (TFTR), two interrupter systems with axial-magnetic-field vacuum interrupters were subjected to 1000 operations each at 24 kA and 25 kV. No failures were observed during these tests,[30] which paved the way for the design and construction of the switching system eventually installed in the ohmic heating circuit of the TFTR.

Another type of vacuum switch is the triggered vacuum gap (TVG) switch. It has fixed electrodes, usually disc or hemispherical, and a trigger system to initiate conduction (Fig. 25). The TVG switch can be triggered to close rapidly, with turn-on times in the microsecond range for commercial devices and less than 100 nanoseconds for special, low-inductance laboratory devices.[31] As long as the discharge remains in the diffuse mode during conduction, triggered vacuum gap switches can be used as current-zero opening switches. With the trigger system, such switches also have a high repetition rate capability. In fact, such operation with commercial vacuum gaps (Fig. 26), has already been demonstrated[1] at a power level of 75 MW (8.6 kA at 8.6 kV) and a pulse repetition rate of 5 kpps.

The record capability demonstrated by the rod array triggered vacuum gap (RATVG) switch should be noted. The RATVG switch was developed by General Electric Company under contract to the Electric Power Research Institute (EPRI) for possible use in ac transmissions systems.[32,33] While rapid advances in the field of SF_6 circuit breakers overshadowed the capabilities of vacuum switches for the intended ac applications, the performance of the RATVG switch is still worth noting for pulsed power applications. The switch consists of rods attached to the anode and cathode, oriented axially and interleaved (Fig. 27). The arc burns in a circumferential direction between the rods. In ac tests at GE, the best of the RATVG switches--the G1 tube--conducted a half-cycle current pulse of 150 kA peak, interrupted at current zero, and recovered to 135 kV within 150 μs. A similar experimental device successfully carried an ac half-cycle current pulse of 240 kA peak with an arc voltage of less than 70 V. Inspection afterwards showed no visible damage of the electrodes. Both visual observations of the arc and records of the switch voltage waveforms confirmed that the vacuum arc remained in a diffuse mode during all of the tests of the rod array structure, including the test at 240 kA peak (personal communication, J. Rich and C. Goody of GE, Dec. 1982). Therefore, the

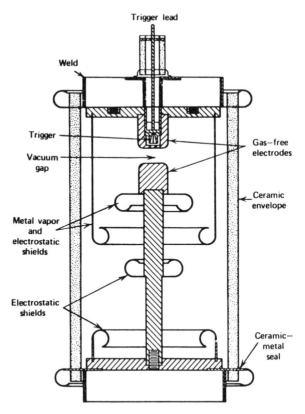

Fig. 25a. Schematic diagram of a typical TVG switch.[23]

RATVG switch should perform nearly as well in dc systems with the fast counterpulse technique. With its trigger system and high power capability, the RATVG switch is a natural candidate for extending the capabilities of repetitive transfer circuits.[1] With the G1 tube, operation in the range of 100 kA, 100 kV, and 1 kpps should be possible.

The history, physics, and development of vacuum switches have been reviewed by Lafferty,[23] Lee,[16] Flurscheim,[17] and Kimblin,[34-36] while a bibliography by Miller[37] covers the years 1897-1982. The first noteworthy research into vacuum switch phenomena was performed by Sorenson and Mendenhall[38] from 1923 to 1926. They achieved the interruption of 1.3 kA peak against 58.8 kV peak. However, early attempts to develop vacuum switch technology did not fare well "because of the lack of supporting technologies in vacuum and metallurgical processing," according to Lafferty.[23] Lee[16] states that the major problems were (1) the evolution of gas from the electrodes during arcing (greatly reducing or destroying the

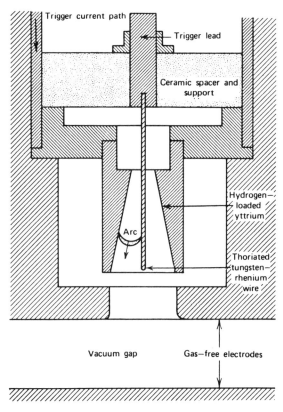

Fig. 25b. Schematic diagram of a typical TVG trigger assembly switch.[23]

current interruption capability), (2) the lack of rugged and reliable metal-to-glass (or ceramic) seals, and (3) the welding of the contacts on closing. The first problem was the most severe, and it was not solved until the mid-1950s when researchers at the General Electric Co. applied the technique of zone refining, borrowed from the semiconductor industry, to obtain gas-free copper electrodes.[39] Rapid commercial development followed, with the first use of vacuum switches in power systems described by Jennings[40] in 1956, and the first 5 years of industrial use reviewed by Ross.[41] By 1960, the ratings had been pushed to very respectable levels,[42] and vacuum interrupter switches with power ratings as high as 250-MVA were in commercial production by 1963.[39]

Over the years, a large body of information on vacuum arcs and switching has been accumulated, as evidenced by the 2752 papers referenced in the bibliography by Miller.[37] Most of the research has been concerned with the physics of the vacuum arc: its initiation, conduction state, interaction with the anode and cathode

Fig. 26 Picture of commercial TVG switches used at Los Alamos National Laboratory.[1]

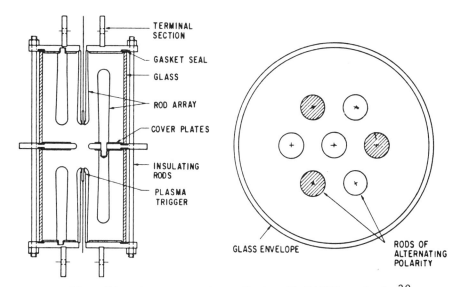

Fig. 27. Cross section of the G1 RATVG switch.[32]

surfaces, and extinction. Other areas of research included electrode erosion and metallurgy, vacuum technology, triggering, magnetic field effects, special devices, and applications.

Once initiated by a trigger pulse or the bridge-point explosion of separating contacts, the vacuum arc is sustained by metal vapor evaporated from the electrode surfaces. Considerable research has been done on various gap triggering methods, including work by Lafferty,[23,43] Farrall,[44] Gilmour,[45] Boxman,[46] and Thompson.[47] For copper electrodes, the metal vapor is about 50 per cent ionized and streams away from many small cathode spots (each carrying about 100 A) at a velocity of about 10^4 m/s.[48-50] The vacuum arc remains in a diffuse mode (with the cathode spots spread uniformly over the surface area of the electrodes) so long as the current does not exceed a given threshold value, determined primarily by electrode metallurgy and geometry, gap separation, and arc history.

The cathode spots have a short thermal time constant, emitting metal vapor for only a few microseconds after the current stops. The interelectrode gap clears rapidly due to the high ejection velocity of the vapor and the vacuum background pressure in the device. The vacuum arc plasma rapidly condenses on the surfaces of the electrodes and vapor shields. Interelectrode gap clearing times as short as 3-5 μs have been observed with copper electrodes.[50] The actual plasma deionization time required depends on a number of variables, including gap spacing, electrode geometry and metallurgy, the current reduction rate, di/dt, before current-zero, and the voltage recovery rate, dv/dt, after interruption. For any given switch and expected operating conditions, the shortest current-zero time that still allows reliable interruption is best determined experimentally. For much of the single-pulse vacuum interrupter work done at Los Alamos National Laboratory,[24,29,51] current-zero times of about 20 μs (following a current fall-time of about 20 μs) were sufficient to allow currents up to 40 kA to be interrupted against peak recovery voltages of 30 kV.

Reliable current-zero opening of vacuum switches requires that the vacuum arc remain in the diffuse mode during conduction. In this mode, the electrode activity occurs almost entirely at the cathode. If the current level exceeds a given threshold value, then the arc enters the constricted mode and heavy anode involvement occurs. The many individual plasma columns present in the diffuse mode coalesce into one or two constricted arcs, each anchored to an anode spot or jet. In this mode, the vacuum arc is highly destructive, causing considerable electrode erosion from each anode jet. Another major problem with the constricted-arc mode is that the hot anode spots continue to supply plasma to the electrode-gap region long after the current has been removed. This delays switch recovery for times on the order of the thermal time constant of the anode spots,[17] at least 100 µs, decreasing the chances for current interruption with a fast current zero. To achieve an adequate current-zero time for constricted-arc conditions, the counterpulse capacitor bank would have to be about 10 times larger than for the fast counterpulse technique used with diffuse arcs. Such a capacitor would be impractical for most applications.

Some work has been done in extending the constricted-arc current threshold and in reducing its damaging effects. For ac systems, special electrode geometries use the arc's own magnetic field to move the arc across the electrode surfaces to prevent the constricted arc from anchoring to one spot on the anode.[17,35] This method reduces the anode thermal time constants and has extended the interruption capability of vacuum switches in ac systems, where the current falltime is about 4 ms. However, this method gives little improvement for dc systems in which the fast counterpulse technique is used.

One method that has proven beneficial in both ac and dc systems is the use of a magnetic field parallel to the direction of the arc, along the centerline axis of the anode and cathode.[36] The magnetic field can be generated either with an external power supply or by the switch current itself. In either case, the magnetic field reduces the magnitude and "roughness" of the arc voltage waveform and increases the constricted-arc current threshold. The axial magnetic field lines imbedded in the plasma help prevent the plasma columns from combining to form a constricted arc. Improvements of 10-100 per cent in current interruption capability were achieved in switches at Los Alamos National Laboratory by using axial magnetic fields.[51] Yanabu used this method to interrupt currents up to 280 kA peak in a single device.[52]

<u>Thyristors</u>. Solid-state thyristor switches, four-layer p-n-p-n semiconductor devices generally made of silicon, operate similarly to vacuum switches and hydrogen thyratrons in that they are triggered closed and then remain conducting until undergoing a current zero. In fact, the thyristor's name stems from its intended application as a "transistor replacement for the thyratron." With a 30-year history, thyristor technology today is mature and well developed, with a large commercial market.[53] Device ratings continue to increase as wafer size and quality increase. A large amount of literature describes the development of thyristors and their various applications.[54-63]

The types of thyristors that are most suitable for fast, high power opening switch applications are the silicon controlled rectifier (SCR) switch and the gate turn-off (GTO) thyristor. As a true current-zero switch, the SCR requires the full device current to be counterpulsed to zero. The GTO thyristor, with a faster turn-off time, can be opened with a current counterpulse, or with a reverse gate pulse of about 20 per cent of the device current, or both. The advantages of such thyristors over vacuum switches are that they have inherent reverse blocking capability, high reliability within rated operation, and no electrode erosion, implying long life. However, they are much more susceptible to damage from out-of-limit operation, even if momentary, and are considerably more expensive.

The SCR switch, the best developed and most common type of thyristor, is a unidirectional device with three terminals--anode, gate, and cathode--as shown in Fig. 28. Within its rated voltage range, the SCR remains nonconducting, exhibiting both forward and reverse blocking. It can be triggered into a highly-conductive state in the forward direction by applying a short trigger pulse to the gate. As suggested by the device circuit symbol (Fig. 28), the SCR can be thought of as a triggered diode, blocking in the reverse direction and not conducting in the forward direction until triggered (unless device breakdown voltages are exceeded).

Figure 29 shows typical SCR voltage-current (V-I) characteristic curves for various values of gate current. In the low-current part of the curve that is nearly symmetrical about the origin, between $-V_{RA}$ and $+V_{BO}$, the switch is in the blocking mode, passing only a small leakage current. Forward current is blocked by the center p-n junction being reverse biased, and reverse current is blocked by the two outer junctions being reverse biased. If the reverse voltage exceeds the reverse avalanche breakdown voltage, $-V_{RA}$, the two outer junctions break down and the SCR switch conducts in the reverse direction in a manner similar to a Zener diode. In the forward direction, the switch blocks conduction until the forward voltage exceeds V_{BO}, the forward breakover voltage for zero gate current. The leakage current increases rapidly with device temperature, causing the total loss of forward blocking[54] if the junction temperature exceeds about 125°C. The leakage also increases with voltage until the forward voltage exceeds V_{BO}, when the leakage current is large enough to cause the center junction to break down. All the switch junctions then become forward

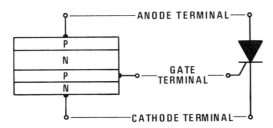

Fig. 28. Circuit schematic symbol of an SCR switch.

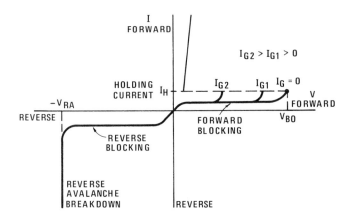

Fig. 29. Volt-ampere characteristic curve of an SCR switch.

biased with the base regions saturated, latching the SCR into the ON-state. In this state, switch operation follows the vertical, low-voltage part of the switch characteristic curve in Fig. 29 in a manner similar to a forward-biased diode. In this region, the switch has a low forward conduction drop, ranging from about 0.8 to 2.5 volts. The conduction drop depends primarily on device construction, increasing with the voltage rating of the device. The switch remains in the high-conduction ON-state as long as current supplied by the external circuit is more than the holding current, I_H, which may be as low as a few mA.

Normally, the SCR switch is triggered into the on state by applying a current pulse to the gate. Gate current flowing across the gate-to-cathode p-n junction has the effect of initiating current flow across the center, reverse biased junction. This in turn adds current to the gate region which again adds current to the center junction. Through this regenerative feedback mechanism, a small gate current pulse of only a few μs duration can switch the SCR on quickly at voltages much less than the forward zero-gate breakover voltage, V_{BO}. Increasing the magnitude of the gate pulse has the effect of decreasing the forward breakover voltage and allowing switching to occur at lower voltages (Fig. 29). With sufficient gate current, the device operates like a simple diode, conducting as soon as forward voltage is applied.

The turn-on switching mechanism is often explained with the help of the two transistor model of the SCR. By separating the four-layer p-n-p-n SCR into two 3-layer devices, one p-n-p and one n-p-n, the device can be modeled as a complimentary pair of transistors, with the SCR gate connection made to the base of the n-p-n transistor, Fig. 30. With zero gate current the base current for each transistor is the same as the collector current for the other

INDUCTIVE ENERGY STORAGE CIRCUITS AND SWITCHES

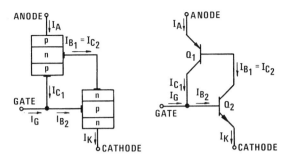

Fig. 30. Two-transistor model of SCR switch.

transistor. Applying a gate pulse supplies current to the base of transistor Q2, turning it on. As the collector current of Q2 increases, it increases the base current of Q1, turning Q1 on. The increase in collector current of Q1 then increases the base current of Q2. The regenerative feedback between Q1 and Q2 operates to turn both devices on quickly, driving them into saturation.

Although the two-transistor model is helpful in understanding the basic mechanisms involved in thyristor operation, it must be used carefully. In particular, it is inadequate to fully explain the device turn-on process. Rather than turn on the entire lower n-p-n region, as indicated with the two-transistor model, the gate current actually turns on only a small region near the gate. This active region must then spread throughout the wafer for the entire unit to be considered switched on. If too much current is passed by the device during the turn-on time, device failure in the region near the gate results. The turn-on time can be decreased by using special gate geometries and/or by driving the gate near its maximum allowable current. Significant improvements in turn-on time are possible with light triggering.[64] In all cases, the rate-of-rise of current, di/dt, during turn on must be limited by the external circuit to a safe level, as specified by the manufacturer for each device.

Opening of the SCR switch requires the external circuit to provide a current zero for a time sufficient to allow device recovery and its return to the blocking state. During the current-zero time, the carriers are swept out of the two outer junctions, but the center junction recovers only through carrier recombination. The turn-off time can be decreased by providing recombination sites or traps (such as gold atoms) in the silicon, but this results in a higher forward voltage drop. Turn off can also be aided by increasing the inverse voltage during recovery as well as by providing a negative gate current. On the other hand, when the thickness of the center junction is increased to provide a higher blocking voltage rating, the turn-off time increases because a larger volume of silicon must recover. The turn-off time also increases as the rate-of-rise of reapplied recovery voltage, dv/dt, increases. If the voltage risetime exceeds the maximum allowable rate, the switch simply fails to open because displacement current through the junction capacitance has the same affect as applying external gate current. Finally, the turn-off time increases rapidly when junction temperatures exceed 125°C. Recovery times for actual devices depend on specific construction details and ratings and range from a few µs to several hundred µs.

The GTO thyristor operates in about the same way as a conventional SCR switch during forward and reverse blocking, during device turn-on, and during current-zero turn-off.[65] The key difference is that the GTO thyristor can also be turned off by simply applying a reverse gate pulse that is only a small fraction of the principal device current.[66] Turn off with reverse gate drive alone has the advantage of isolating the turn-off driving circuit, but the switch then operates like a direct-interruption switch with the attendant problem of having to dissipate the energy associated with any stray inductances as discussed earlier. An RC snubber network and/or transient voltage suppressor should be placed across the GTO switch to absorb this energy. Operation of the GTO switch as a current-zero switch is essentially the same as with SCR switches, except that the turn-off time can be reduced considerably with reverse gate drive. Therefore, the remainder of this discussion will be limited to SCRs.

The design and manufacture of thyristor switches often involve tradeoffs between various device capabilities such as voltage, current, turn-off time, turn-on di/dt, and turn-off dv/dt. In general, device speed is obtained at the expense of power handling capability and vice versa, while both can be improved by increasing the silicon area.[57] An exception is the use of light triggering to greatly reduce turn-on time without sacrificing power ratings.

The voltage ratings of thyristors are given for both steady state and transient operation and for forward and reverse polarities.[54-57] The voltage ratings depend mostly on the resistivity and thickness of the silicon layers. In the forward direction, the voltage-blocking capability is also temperature sensitive. An increase in the junction temperature above the rated value reduces the switch's reliability and adversely affects its switching characteristics. The switch voltage rating should be about 2.5 times the peak circuit voltage, including transients. This can be reduced to a safety margin of about 1.7 if transient suppressors, such as varistors, are used.[54]

Current ratings are established for thyristors to prevent burn out on short and long time scales.[54-57] The peak current that can be carried in a short pulse depends on the silicon's maximum allow-

INDUCTIVE ENERGY STORAGE CIRCUITS AND SWITCHES

able current density and its upper temperature limit and can be 10-15 times the steady-state current rating. Short-pulse current limits are given by the 1/2-cycle, 3-cycle, and 10-cycle (60-Hz ac) ratings. Steady-state current limits are determined by the power dissipation within the switch and how effectively the waste heat is carried away by the device structure and its heat sink. If the switch is allowed to heat up past its maximum operating temperature, usually about 125°C, the turn-off and voltage blocking capabilities will be severely impaired. Steady-state limits are given in terms of both the average current and the RMS (effective dc) current. Device specifications from the manufacturer give limits for both sinusoidal and rectangular waveforms. If the SCR current is neither of these, the waveform must be approximated by an equivalent combination of rectangular waveforms when calculating the steady-state limits.

As mentioned previously, excessive di/dt during turn on and dv/dt during turn off can both cause device failure. Therefore, limits on these parameters are also specified for each device. Conventional SCRs are seldom rated above 1000 A/μs or above 1000 V/μs, although an experimental light-triggered thyristor[64] has demonstrated a turn-on di/dt capability of at least 40,000 A/μs. If necessary to maintain operation within the specified limits, a series saturable reactor can be used to limit the flow of switch current until the device is fully turned on, and an RC snubber circuit can be placed in parallel with the device to control the recovery voltage rise, to be described in more detail later.

Because of speed-versus-power design tradeoffs, SCR switches are usually manufactured as one of two types, phase control or inverter, with the turn-off time their major difference. Generally, phase control SCRs have turn-off times ranging from 250-700 μs while inverter SCRs have turn-off times ranging from 40-120 μs.[67,68] The turn-off time usually increases with increasing device size and voltage rating.[69] The faster turn-off times in inverter SCRs are obtained at the expense of a higher forward voltage drop, a reduced current rating, and a lower forward breakdown voltage for the same sized device.

Medium-power SCRs are generally stud-mounted while high-power SCRs are supplied in a ceramic disk which allows device cooling from both ends. The diameters of commonly used disk packages in inches (and the associated silicon wafers in mm) are: 2 inch (33-38 mm), 3 inch (50 mm), 4 inch (67 mm), 5 inch (75-83 mm), and 6 inch (100 mm).

Typical phase control SCRs available include a 67-mm device[67] rated at 1400 A_{avg}/4000 V and 83-mm devices[68] rated at 1300 A_{avg}/5500 V or 1830 A_{avg}/4400 V. Typical inverter SCRs include a 38-mm device rated at 440 A_{avg}/2000 V, a 50-mm device[68] rated at

780 A_{avg}/2500 V, and a 67-mm device[67] rated at 1400 A_{avg}/1800 V. Thyristors with a 100-mm wafer diameter are also commercially available, primarily for HVDC converters, and have steady-state current ratings[68] up to 2570 A_{avg} and blocking voltages up to 4400 V.

Many high-power switch applications have voltage and current requirements greatly exceeding the ratings of single thyristors. In these cases, a composite switch can be assembled by stacking individual thyristors in a series/parallel array--thereby adding the voltage capability of units in series and the current capability of units in parallel. Extra precautions must be taken when building such a composite switch--a thyristor stack--to insure that the voltage is distributed equally across series units (by using voltage balancing networks) and that the current is distributed equally among parallel switch sections (by using current balancing resistors and/or transformers). Differences in the turn-on di/dt and turn-off dv/dt of individual devices require a derating of the switch array to insure safe operation of each element.[70] Such assemblies are most commonly used in HVDC transmission systems.[69-72]

Circuit Considerations

Size of Counterpulse Capacitor

To provide adequate current-zero conditions for switch recovery, the counterpulse capacitor must both reduce the opening switch current to zero and then supply a current equal to the initial switch current during the entire current-zero time, Fig. 31. Therefore, assuming a linear reduction in the switch current during the fall time, the capacitor must supply a total charge q_0 given by

$$q_0 = I_0(0.5T_{Fall} + T_{I\text{-}Zero}), \tag{27}$$

where I_0 is the current flowing in the opening switch just prior to the counterpulse operation, T_{Fall} is the fall time of the switch current from I_0 to zero, and $T_{I\text{-}Zero}$ is the switch current-zero time. Now the charge on a capacitor C is given by

$$q_0 = CV_0, \tag{28}$$

where V_0 is the capacitor voltage. Combining Eqs. (27) and (28) gives the counterpulse capacitor design equation

$$CV_0 = I_0(0.5T_{Fall} + T_{I\text{-}Zero}). \tag{29}$$

In this expression, T_{Fall} is determined by the parameters of the counterpulse circuit and $T_{I\text{-}Zero}$ is fixed by the recovery time of the particular current-zero opening switch used. Equation (29) then fixes the product of CV_0, the total initial charge q_0 needed. If the circuit designer has some freedom in the choice of the initial capacitor voltage, there will be a corresponding range of acceptable capacitor values. For repetitive applications, the initial capacitor voltage is often chosen to be about equal to the expected load voltage, and Eq. (29) then determines the capacitor size accordingly.

Repetition Rate Limit.[73] After the switch current has been forced to zero, the full initial current, I_0, flows through the counterpulse capacitor (Fig. 31). This causes the magnitude of the capacitor voltage to decrease linearly during the switch current-zero time, T_{I-Zero}, and to increase linearly during the rise time, T_{Rise}. For repetitive applications, the capacitor must always be recharged to the same voltage it had initially so that it will have sufficient charge for each new counterpulse operation. The recharge condition can then be expressed as

$$q_R = I_0 T_{Rise}, \tag{30}$$

where q_R is the capacitor charge replenished during T_{Rise}. For symmetrical recharge, $q_R = q_0$, and combining Eqs. (27) and (30) gives

$$I_0 T_{Rise} = I_0 (0.5 T_{Fall} + T_{I-Zero}). \tag{31}$$

Dividing by I_0 then gives

$$T_{Rise} = 0.5 T_{Fall} + T_{I-Zero}. \tag{32}$$

Now the total time, T_{CP}, involved in one counterpulse operation is simply the sum of the switch fall time, the switch current-zero time, and the capacitor recharge time, and is given by

$$T_{CP} = T_{Fall} + T_{I-Zero} + T_{Rise}. \tag{33}$$

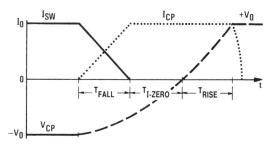

Fig. 31. Waveforms of switch current and capacitor voltage during typical counterpulse operation.

For conditions of symmetrical recharge, with T_{Rise} given by Eq. (32), the total counterpulse time becomes

$$T_{CP} = 1.5 T_{Fall} + 2 T_{I-Zero}, \qquad (34)$$

and the maximum pulse repetition rate, PRR_{max}, is given by

$$PRR_{max} = 1/T_{CP}. \qquad (35)$$

For example, if one assumes typical numbers for vacuum interrupters,

$$T_{Fall} = T_{I-Zero} = 20\ \mu s, \qquad (36)$$

then the total counterpulse time is 70 μs and the pulse repetition rate cannot exceed about 14 kHz.

Load Pulse Risetime.[1] During the counterpulse operation described above, the load is usually isolated from the rest of circuit. Once the counterpulse capacitor has been recharged, the load pulse is initiated by closing the load isolation switch. The equivalent circuit for an inductive energy storage circuit at the instant of load switch closing is shown in Fig. 32. By this time in the transfer cycle, the main opening switch (not shown in Fig. 32) has already recovered so that the storage inductor current, I_1, flows through the capacitor, C, recharging it to a voltage, V_C, just prior to load switch closing. It is convenient to relate the capacitor voltage to the resistive component of the load voltage, $I_1 R$, through the charge factor, K, as

$$K = V_C/I_1 R. \qquad (37)$$

The equivalent circuit just described, with the storage inductor L_1 assumed to be much larger than the load inductor L_2, was

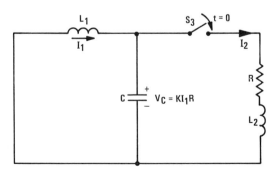

Fig. 32. Equivalent circuit diagram of counter-pulsed inductive storage circuit at the instant the load switch is closed.

analyzed to determine the nature of the output pulse when using a current-zero switch in series with the counterpulse capacitor[1]. Depending on the relative values of R, L_2, and C, the solutions for the load current waveforms are either underdamped ($\alpha < \omega_o$), critically damped ($\alpha = \omega_o$), or overdamped ($\alpha > \omega_o$), where the system damping factor α is

$$\alpha = R/2L_2 \tag{38}$$

and the natural (undamped) system frequency ω_o is

$$\omega_o = (1/L_2C)^{1/2}. \tag{39}$$

Complete analytical solutions for the load current waveforms are derived in Ref. 1. Figure 33 shows representative waveforms for the case of K = 1. The load current, I_2, is normalized with respect to the initial storage coil current, I_1, while the time, t, is normalized with respect to the zero-damped case risetime, t_{RZ}, given by

$$t_{RZ} = \pi/2\ \omega_o. \tag{40}$$

When the load pulse magnitude equals the storage current in L_1, the counterpulse capacitor experiences a current zero as its current reverses. The dotted portions of the current waveforms in Fig. 33 show the load current excursion if the capacitor current reverses and it discharges into the load. In repetitive applica-

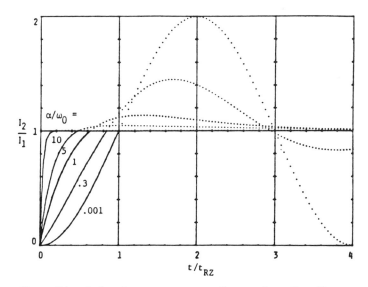

Fig. 33. Normalized load current waveforms for the three general cases of damping (with K = 1).[1]

tions, the capacitive energy must be saved to generate the next counterpulse operation, and a current-zero switch must be placed in series with the capacitor to prevent its current from reversing. With a current-zero counterpulse switch, the load pulse follows the solid line portions of the waveforms shown in Fig. 33 and is clipped at a peak value of $I_2 = I_1$. Therefore, this type of circuit produces a fast-rising load pulse with a flat top.

The rise time for the "clipped" case, where the counterpulse switch prevents capacitor current reversal and load pulse overshoot, was analyzed in Ref. 1 and analytical solutions obtained for the three cases of damping and various values of K, as shown in Fig. 34. The 0-to-100 per cent rise time, t_R, is normalized with respect to the zero-damped risetime, t_{RZ}, as already given by Eq. (40). The dotted curve for the critically-damped case separates the curves for the underdamped case from the curves for the overdamped case. If the circuit is very underdamped, then the risetime does not depend much on the charge factor K and the risetime is about the same as t_{RZ}. As damping increases, an increase in K resulting from an increase in capacitor voltage relative to the expected load voltage serves to decrease the load pulse risetime. When the circuit is overdamped, the risetime depends strongly on K. In fact, for strong overdamping, the counterpulse capacitor must be charged to a minimum value (with K about one or greater) for the output current pulse to ever reach the same magnitude as the coil current.

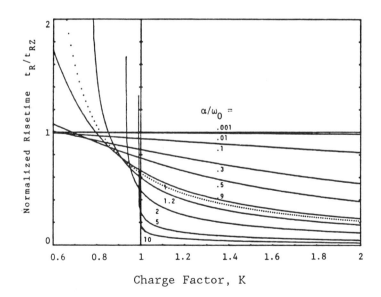

Fig. 34. Plots of the "clipped-case" normalized rise times of the output current pulses as a function of the charge factor, K and the damping ration α/ω_0.[1]

This analysis shows that the output risetime and the counterpulse time depend on substantially different conditions: the risetime is a function of the output discharge loop conditions while the counterpulse depends largely on the opening switch recovery time. For example, in a 75-MW, 5-kpps repetitive pulse demonstration[1] output risetimes of 4 μs (0-100 per cent) were achieved in a circuit where the total counterpulse time was about 70 μs.

Device Protection Circuits. The selection of current-zero switches for a particular application is based primarily on their ratings of peak blocking voltage and conduction current (either steady-state or pulsed). However, careful attention must also be given to transient circuit conditions and their effect on the switch. Solid-state switches are much more susceptible to momentary overloads of voltage or current than are vacuum switches and they must be protected accordingly. A general design guide for thyristors is to use devices with a voltage rating about 2.5 times the expected peak system voltage. This can be reduced to about 1.7 if fast transient voltage suppressors, such as varistors, are installed in parallel with the switches.[59] Likewise, a good design procedure to protect against damaging current surges in solid-state switches is to install a series fuse with a rating less than the switch's I^2t rating.[59]

The transient conditions occurring during device turn on and turn off must also be addressed. Excessive di/dt during turn on can easily damage a thyristor in the region near the gate, while excessive dv/dt during turn off can prevent switch recovery. A standard procedure to keep the turn-on current within di/dt limits is to add an inductor, usually in the form of a saturable reactor, in series with the switch. The volt-second rating of the reactor is selected to keep the switch current low until the device is fully turned on. After the reactor saturates, the full system current is passed through the switch.[54,74] The same reactor also provides assistance during turn off, providing a current-zero time which depends on the volt-second rating of the reactor and the average inverse voltage on the counterpulse capacitor during the current-zero time.[1,75,76]

Immediately after the switch opens electrically, the switch is stressed by a recovery voltage which depends on the makeup of the rest of the circuit. Complete and sustained switch opening requires that not only the recovery voltage's peak value but also its rate of rise remain well within ratings. In an inductive storage circuit with a current-zero opening switch, for example, the switch voltage is limited to a rate of rise given by

$$dv/dt = I_0/C. \tag{41}$$

where I_0 is the value of the switch current interrupted and C is the value of the counterpulse capacitor. If the counterpulse capacitor is too small to keep dv/dt within limits, or in circuits without such a capacitor, it is common to install a snubber circuit consisting of a series resistor-capacitor combination in parallel with the switch. The capacitor limits the voltage rise, as described above, and the resistor is used to keep the capacitor discharge within the switch's turn-on di/dt limits.[77]

Application Example

Current-zero switches have found commercial application in HVDC transmission systems and in high power converter-inverters. In the pulsed-power field, they have been used to generate a 75-MW pulse train at a 5-kpps rate using an inductive energy store as shown in Fig. 35. The repetitive pulses of 8.6 kA peak were generated using the counterpulse technique with a vacuum interrupter and triggered vacuum gaps.[1] The pulses were 27 µs wide and had a risetime of only 4 µs, even though the counterpulse time was about 70 µs. A comparison of the total energy transferred to the load versus the decrease in stored energy during the pulse train showed an operational efficiency of 77 per cent. The high operational efficiency can be attributed to the advantages of using a capacitive transfer circuit and the counterpulse technique, in which switches open under conditions of zero current.

Acknowledgments

The author gratefully acknowledges the long-term guidance and encouragement provided by Dr. Magne Kristiansen of Texas Tech University as a teacher, mentor, and friend during the last 20 years. It would be impossible to name all the others who have provided valuable feedback during the course of this research, but the help of some key individuals is highlighted in Ref. 1. In addition the author would like to thank Dr. Heinrich Boenig of Los Alamos, Dr. Stephen Kuznetsov of Westinghouse, and Mr. R.L. Cummins of Brown Boveri Inc. for helpful discussions concerning thyristors. Finally, the Los Alamos National Laboratory should be recognized for providing the funding and working environment to make all this possible.

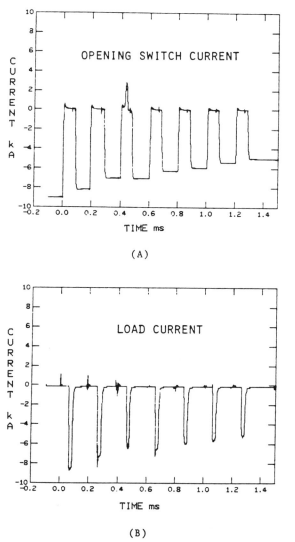

Fig. 35. Waveforms of the opening switch current (a) and the load current (b) during 75-MW, 5-kpps pulse train operation, of an inductive storage circuit[1].

REFERENCES

1. E.M. Honig, "Repetitive Energy Transfers From an Inductive Energy Store" (Ph.D. Dissertation, Texas Tech University, Aug. 1984), University Microfilms Int. Pub. No. 84-27984. Copyright © 1984 Emanuel M. Honig, Los Alamos, NM. Also available as Los Alamos National Laboratory Report No. LA-10238-T (1985).
2. R.F. Caristi and D.V. Turnquist, "Repetitive Series Interrupter II, Final Report," U.S. Army ERADCOM Research and Development Technical Report No. DELET-TR-76-1301-F (1979).
3. H.C. Early and F.J. Martin, Method of Producing a Fast Current Rise from Energy Storage Capacitors, Rev. Sci. Instr., 36:1000 (1965).
4. R.E. Reinovsky, W.L. Baker, Y.G. Chen, J. Holmes, and E.A. Lopez, SHIVA STAR Inductive Pulse Compression System, 4th IEEE Pulsed Power Conf., IEEE Cat. No. 83CH1908-3 (1983).
5. S.A. Nasar and H.H. Woodson, Storage and Transfer of Energy for Pulsed-Power Applications, 6th Symp. Engr. Problems Fusion Research, IEEE Pub. No. 75CH1097-5-NPS (1975).
6. M.F. Rose, Techniques and Applications of Pulsed Power Technology, 16th Intersociety Energy Conversion Engineering Conf. (1981).
7. M. Kristiansen, Editor, Proceedings of the Workshop on Switching Requirements and R & D for Fusion Reactors, Electric Power Research Institute Report No. ER-376-SR (1977).
8. M. Kristiansen and K.H. Schoenbach, Editors, US Army Workshop on Repetitive Opening Switches, DTIC No. AD-A110770 (1981).
9. M. Kristiansen and K.H. Schoenbach, Editors, US Army Workshop on Diffuse Discharge Opening Switches, DTIC No. AD-A115883 (1982).
10. W.M. Portnoy and M. Kristiansen, Editors, US Army Workshop on Solid State Switches for Pulsed Power, DTIC No. AD-A132687 (1983).
11. M.O. Hagler and M. Kristiansen, Editors, U.S. Army Workshop on Repetitive Spark Gap Operation, DTIC No. AD-A132688 (1983).
12. H.C. Early, Principles of Inductive Energy Storage, Study S-104: IDA Pulse-Power Conf., Vol. I, Report No. IDA/HQ63-1412 (1963).
13. H.L. Laquer, Superconductivity, Energy Storage, and Switching, Int. Conf. on Energy Storage, Compression, and Switching (1974).
14. S.L. Wipf, Superconductive Energy Storage, NASA Technical Translation No. NASA-TT-F-15109 (N73-31676) (1973).

15. T.F. Trost, P.E. Garrison, and T.R. Burkes, Pulse Power Systems Employing Inductive Energy Storage, <u>1st IEEE Int. Pulsed Power Conf.</u>, IEEE Pub. No. 76CH1147-8-REG5 (1976).
16. T.H. Lee, "Physics and Engineering of High Power Switching Devices", MIT Press, Cambridge, MA (1975).
17. C.H. Flurscheim, Editor, "Power Circuit Breaker Theory and Design", Peter Peregrinus Ltd., Stevenage, England (1975).
18. T.R. Burkes, J.P. Craig, M. O. Hagler, M. Kristiansen, and W.M. Portnoy, A Review of High-Power Switch Technology, <u>IEEE Trans. Electron Devices</u>, ED-26:1401 (1979).
19. K.H. Schoenbach, M. Kristiansen, and G. Schaefer, A Review of Opening Switch Technology for Inductive Energy Storage, <u>Proc. IEEE</u>, 72:1019 (1984).
20. R. Carruthers, Energy Storage for Thermonuclear Research, <u>Proc. IEE, Part A Supplement 2</u>, 106:166 (1959).
21. E.M. Honig, Progress in Developing Repetitive Pulse Systems Utilizing Inductive Energy Storage, <u>4th IEEE Pulsed Power Conf.</u>, IEEE Pub. No. 83CH1908-3 (1983).
22. H.H. Woodson and W.F. Weldon, Energy Considerations in Switching Current From an Inductive Store into a Railgun, <u>4th IEEE Pulsed Power Conf.</u>, IEEE Pub. No. 83CH1908-3 (1983).
23. J.M. Lafferty, Editor, "Vacuum Arcs, Theory and Application", John Wiley & Sons, New York, NY (1980).
24. E.M. Honig, C.E. Swannack, R.W. Warren, and D.H. Whitaker, Progress in Switching Technology for METS Systems, <u>IEEE Trans. Plasma Science</u>, PS-5:61 (1977).
25. R.W. Warren and E.M. Honig, The Use of Vacuum Interrupters at Very High Currents, <u>Proc. 13th Pulse Power Modulator Symp.</u>, IEEE Pub. No. 78CH1371-4-ED (1978).
26. E.M. Honig and R.W. Warren, The Use of Vacuum Interrupters and Bypass Switches to Carry Currents for Long Times, <u>Proc. 13th Pulse Power Modulator Symp.</u>, IEEE Pub. No. 78CH1371-4-ED (1978).
27. W.M. Parsons, Vacuum Switchgear for Fusion Experiments, <u>Physica</u> 104C:161 (1981).
28. R.W. Warren, The Early Counterpulse Technique Applied to Vacuum Interrupters, <u>Los Alamos National Laboratory Report</u> No. LA-8108-MS (1979).
29. R. Warren, M. Parsons, E. Honig, and J. Lindsay, Tests of Vacuum Interrupters for the Tokamak Fusion Test Reactor, <u>Los Alamos Scientific Laboratory Report</u> No. LA-7759-MS (1979).
30. P. Bellomo, J. Calpin, R. Cassel, and H. Zuvers, Plasma Striking Voltage Production, <u>Proc. 9th Symp. Engr. Problems Fusion Research</u>, IEEE Pub. No. 81CH1715-2-NPS (1981).

31. F.T. Warren, J.M. Wilson, J.E. Thompson, R.L. Boxman, and T.S. Sudarshan, Vacuum Switch Trigger Delay Characteristics, IEEE Trans. Plasma Science, PS-10:298 (1982).
32. J.A. Rich, C.P. Goody, and J.C. Sofianek, High Power Triggered Vacuum Gap of Rod Array Type, General Electric Report No. 81CRD321 (1981).
33. J.A. Rich, G.A. Farrall, I. Imam, and J.C. Sofianek, Development of a High-Power Vacuum Interrupter, Electric Power Research Institute Report No. EL-1895 (1981).
34. C.W. Kimblin, P.G. Slade, J.G. Gorman, and R.E. Voshall, Vacuum Interrupters Applied to Pulse Power Switching, 3rd IEEE Int. Pulsed Power Conf., IEEE Pub. No. 81CH1662-6 (1981).
35. C.W. Kimblin, Arcing and Interruption Phenomena in AC Vacuum Switchgear and in DC Switches Subjected to Magnetic Fields, IEEE Trans. Plasma Science, PS-11:173 (1983).
36. C.W. Kimblin, A Review of Arcing Phenomena in Vacuum and in the Transition to Atmospheric Pressures, IEEE Trans. Plasma Science, PS-10:322 (1982).
37. H.C. Miller, A Bibliography and Author Index For Electrical Discharges in Vacuum (1897-1982), General Electric Company Technical Information Series Report No. GEPP-TIS-366c (1984).
38. R.W. Sorensen and H.E. Mendenhall, Vacuum Switching Experiments at California Institute of Technology, AIEE Trans., Pt. III, 45:1203 (1926).
39. J.D. Cobine, Research and Development Leading to High-Power Vacuum Interrupters, Electrical Engineering, 81:13 (1962).
40. J.E. Jennings, A.C. Schwager, and H.C. Ross, Vacuum Switches for Power Systems, AIEE Trans., Pt. III, 75:462 (1956).
41. H.C. Ross, Vacuum Power Switches: 5 Years of Field Application and Testing, AIEE Trans., Pt. III, 80:758 (1961).
42. T.H. Lee, A. Greenwood, D.W. Crouch, and C.H. Titus, Development of Power Vacuum Interrupters, AIEE Trans., Pt. III, 81:629 (1963).
43. J.M. Lafferty, Triggered Vacuum Gaps, Proc. IEEE, 54:23 (1966).
44. G.A. Farrall, Vacuum Arcs and Switching, Proc. IEEE, 61:1113 (1973).
45. A.S. Gilmour, and D.L. Lockwood, Pulsed Metallic-Plasma Generators, Proc. IEEE, 60:977 (1972).
46. R.L. Boxman, Triggering Mechanisms in Triggered Vacuum Gaps, IEEE Trans. Electron Devices, ED-24:122 (1977).
47. J.E. Thompson, R.G. Fellers, T.S. Sudarshan, and F.T. Warren, Design of a Triggered Vacuum Gap for Crowbar Operation, Proc. 14th Pulse Power Modulator Symp., IEEE Pub. No. 80CH1573-5-ED (1980).
48. C.W. Kimblin, Erosion and Ionization in the Cathode Spot Regions of Vacuum Arcs, J. Appl. Phys., 44:3074 (1973).

49. M.P. Reece, The Vacuum Switch: Part 1. Properties of the Vacuum Arc, *Proc. IEE*, 110:793 (1963).
50. A. Selzer, Switching in Vacuum: A Review, *IEEE Spectrum*, 8:26 (1971).
51. R.W. Warren, Experiments with Vacuum Interrupters Used for Large DC-Current Interruption, *Los Alamos Scientific Laboratory Report* No. LA-6909-MS (1977).
52. S. Yanabu, E. Kaneko, H. Okumura, and T. Aiyoshi, Novel Electrode Structure of Vacuum Interrupter and Its Practical Application, *IEEE PES Summer Mtg.*, Paper No. 80SM700-5 (1980).
53. J.L. Moll, M. Tanenbaum, J.M. Goldey, and N. Holonyak, P-N-P-N Transistor Switches, *Proc. IRE*, 44:1174 (1956).
54. R.G. Hoft, Editor, "SCR Applications Handbook," International Rectifier Corp., El Segundo, CA (1974).
55. D.R. Grafham and J.C. Hey, Editor, "SCR Manual," General Electric Co., Syracuse, NY (1972).
56. L.R. Rice, Editor, "SCR Designers Handbook," Westinghouse Electric Corp., Youngwood, PA (1970).
57. J.W. Motto, "Introduction to Solid State Power Electronics," Westinghouse Electric Corp., Youngwood, PA (1977).
58. F.E. Gentry, F.W. Gutzwiller, N. Holonyak, and E. Von Zastrow, "Semiconductor Controlled Rectifiers," Prentice-Hall, Inc., Englewood Cliffs, NJ (1964).
59. A. Blicher, "Thyristor Physics," Springer-Verlag, New York Inc., New York, NY (1976).
60. R. Sittig and P. Roggwiller, "Semiconductor Devices for Power Conditioning", Plenum Press, New York, NY, (1982).
61. S.K. Ghandhi, "Semiconductor Power Devices," John Wiley and Sons, Inc., New York, NY (1977).
62. G.T. Coate and L.R. Swain, "High-Power Semiconductor-Magnetic Pulse Generators", MIT Press, Cambridge, MA (1966).
63. J.D. Harnden and F.B. Golden, "Power Semiconductor Applications," Vol. 1 and 2, IEEE Press, New York, NY (1972).
64. L.R. Lowry, Optically Activated Switch, *Air Force Aero Propulsion Laboratory Report* No. AFAPL-TR-78-17, (1978).
65. T. Nagano, T. Yatsuo, and M. Okamura, Characteristics of a 3000-V 1000-A Gate Turn-off Thyristor, *IEEE-IAS '81 Ann. Meet. Rec.* (1981).
66. T. Yatsuo, T. Nagano, H. Fukui, M. Okamura, and S. Sakurada, Ultrahigh-Voltage High-Current Gate Turn-Off Thyristors, *IEEE Trans. Electron Devices*, ED-31:1681 (1984).
67. G.A. Sutherland, "Power Semiconductor Data Book," Westinghouse Electric Corp., Youngwood, PA (1982).
68. "Quick Reference Catalog," *Pub. No. CH-E 3.40733.9 E*, BBC Brown Boveri, Lenzburg, Switzerland (1985).
69. G. Kaplan, Thyristors: Future Workhorses in Power Transmission, *IEEE Spectrum*, 19:40 (1982).

70. J.P. O'Loughlin, Survey of Solid State Switch Applications, Proc. Workshop on Solid State Switches for Pulsed Power, W. M. Portnoy and M. Kristiansen, Ed., DTIC Report No. AD-A132687 (1983).
71. L. Carlsson, Recent Developments in HVDC Convertor Station Design, IEEE Trans. Pwr. App. and Systems, PAS-103:2166 (1984).
72. S. Kobayashi, Y. Tanoue, H. Ikegame, T. Matushita, and Y. Sato, Development of Large-Scale Thyristor DC Circuit Breaker, Proc. 9th Symp. Engin. Problems of Fusion Res., IEEE Pub. No. 81CH1715-2-NPS (1981).
73. E.M. Honig, Vacuum Interrupters and Thyratrons as Opening Switches, Proc. Workshop on Repetitive Opening Switches, M. Kristiansen and K.H. Schoenbach, Editors, DTIC Report No. AD-A110770 (1981).
74. D.A. Paice and P. Wood, Nonlinear Reactors as Protective Elements for Thyristor Circuits, IEEE Trans. Magnetics, MAG-3:228 (1967).
75. A.N. Greenwood and T.H. Lee, Theory and Application of the Commutation Principle for HVDC Circuit Breakers, IEEE Trans. Pwr. App. and Systems, PAS-91:1570 (1972).
76. A.N. Greenwood, "Circuit Interrupting Means for a High Voltage D.C. Circuit," U.S. Patent No. 3,390,305 (1968).
77. W. McMurray, Optimum Snubbers for Power Semiconductors, IEEE Trans. Industry and General Appl., IGA-8:593 (1972).

DIFFUSE DISCHARGE OPENING SWITCHES

K. H. Schoenbach* and G. Schaefer**

*Department of Electrical Engineering
Old Dominion University
Norfolk, Virginia 23508

**Department of Electrical Engineering
Texas Tech University
Lubbock, TX 79409
and Polytechnic University
Farmingdale, NY 11735

INTRODUCTION

In a "Diffuse Discharge Opening Switch," the switch medium is an externally sustained discharge. Most of the research on externally sustained, high pressure (> 1 atm), diffuse discharges has been performed with respect to its application for pumping molecular lasers. The most common sustainment method, the electron-beam (e-beam) controlled discharge technique, was devised and implemented by Daugherty, et. al.,[1] to overcome thermal instabilities in large volume, high energy, CO_2 laser discharges. The potential use of diffuse discharges as fast closing and opening switches was proposed and discussed five years later by Hunter[2] and Koval'chuk and Mesyats.[3,4]

The schematic diagram of a typical externally controlled opening switch as part of an inductive energy storage system is shown in Fig. 1. When the gas between the electrodes is ionized by an e-beam or by some radiation, it becomes conductive. The diffuse discharge switch closes and allows charging of the inductor. During conduction, the discharge is sustained by external means. The reduced electric field strength, E/N, is kept in a range where ionization through the discharge electrons is negligible. When the external ionizing source is turned off, electron attachment and recombination processes in the gas cause the conductivity to decrease and the switch opens. Consequently, the current through the inductor is commutated into the load, and the voltage across the load increases according to the impedance ratio of the load and the storage system.

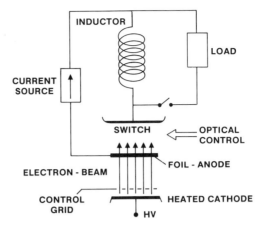

Fig. 1. Schematic of an externally controlled opening switch as part of an inductive energy storage system.

The most energy efficient method at this time to provide for plasma conductivity is the e-beam controlled discharge technique. Up to 50% of the beam energy can be converted directly into ionization energy.[5] The remaining energy mainly goes into excitation. The ionization efficiency can be increased further by using Penning gas mixtures, where part of the excitation energy is converted into ionization through Penning collisions.[6] The way to reduce the conductivity, i.e., to open the switch, is to turn off the e-beam. Optical control of diffuse discharges, on the other hand, allows changing of the conductivity in either direction.[7] By illuminating a discharge with radiation in the wavelength range corresponding to an atomic or molecular transition, the electrical properties of the plasma can be modified through optogalvanic effects.[8] An attractive concept for an externally controlled diffuse discharge switch is to combine the advantages of both the e-beam and optical control to create an e-beam sustained, optically assisted switch.[9]

DISCHARGE ANALYSIS

The goal of a discharge analysis of an externally sustained discharge for switching applications is to evaluate the time dependent impedance of the discharge for a given time dependent electron source in a given circuit. Such analysis will allow the optimization of gas mixtures and other operating conditions. A complete discharge model must consider the bulk of the discharge and fall regions, especially the cathode sheath. It must provide for steady state solutions as well as transient behavior, and should include a stability analysis of the discharge. To compute all this information for all these discharge properties would require a multidimensional, time-dependent model. "It is generally agreed that no single code could, or should, contain all the processes of importance or all possible dimensions, but that more can be learned by analyzing important aspects of the problem with small, specialized and interactive models".[10] A large number of codes for certain aspects exist, developed mainly to provide information for the operation of electron-beam sustained discharges for CO_2 and rare gas halide lasers, which is also relevant to applications as opening switches. Nighan and Wiegand,[11] for example, analyzed the influence of attachers on the steady state properties and striations in molecular gas discharges. Lowke and Davies[12] presented a one-dimensional, steady state analysis including cathode and anode sheaths. (The analysis of the sheaths and instabilities are discussed further in the next section.)

Model calculations aimed toward discharge applications for opening switches have concentrated mainly on the bulk of the discharge (zero-dimensional), on the influence of the proposed optimum gas properties (E/N-dependence of attachment coefficient and mobility), and on the transient behavior (time dependence) of the discharge in an inductive energy storage circuit.[9,13-16] Bulk discharge models are based on solving a set of rate equations for the charged particles and those excited particles which contribute to ionization. The steady state solution of the set of rate equations gives the (V-I)-characteristics of the discharge. For the evaluation of its transient behavior in a given circuit, the set of time dependent rate equations and the circuit equation must be solved simultaneously.

A simplified model of a spatially homogeneous discharge considers only the charged particles.[15] All ionization processes by the discharge electrons are neglected and two-body dissociative attachment is considered to be the dominant attachment process. The rate equations are then given as:

$$\frac{dn_e}{dt} = S - k_{re} n_+ n_e - k_a N_a n_e \tag{1}$$

$$\frac{dn_+}{dt} = S - k_{re} n_+ n_e - k_{ri} n_+ n_- \tag{2}$$

$$\frac{dn_-}{dt} = k_a N_a n_e - k_{ri} n_+ n_- , \tag{3}$$

where k_{re} and k_{ri} are the electron-ion and ion-ion recombination rate coefficients, respectively, k_a is the attachment rate coefficient, N_a is the attacher concentration in the gas mixture, and S is the source function(rate of electron-ion pair production by the e-beam).

Assuming charge neutrality in the discharge so that $n_+ = n_e + n_-$, and with the assumption that $k_{re} = k_{ri} = k_r$, the rate equation can be solved for the <u>steady state</u> electron density, n_{eo}:

$$n_{eo} = \frac{S}{k_a N_a + \sqrt{S k_r}} \quad (4)$$

For electrons being the dominant current carriers, the conductivity is given as:

$$\sigma = \frac{e \mu S}{k_a N_a + \sqrt{S k_r}}, \quad (5)$$

where μ is the electron mobility. This solution reduces to the attachment dominated solution for $k_a N_a \gg k_r n_+$:

$$\sigma_a = \frac{e \mu S}{k_a N_a}, \quad (6)$$

and the recombination dominated solution for $k_a N_a \ll k_r n_+$:

$$\sigma_r = e \mu \sqrt{\frac{S}{k_r}}. \quad (7)$$

Since the diffuse discharge is considered to be used as a switch, that means as an element which controls the energy flow from an energy storage into a load, the power dissipated in the discharge should ideally be negligible compared to the power commutated to the load. This requires a high condutivity at a low reduced field strength, E/N. The desirable gas properties for the conduction phase are therefore, according to Eq. (5), a large electron mobility and small attachment and recombination rate coefficients at low E/N.

In order to minimize losses during the transition from the conductive to the nonconductive state (opening phase) the switch conductance should decay as fast as possible. The decay of the charge densities, after the external ionization source S is turned off, can be calculated by integrating the rate equations (1) and (2), assuming charge neutrality, $n_+ = n_- + n_e$, and constancy of k_r and k_a with respect to changes in field strength, E/N, and correspondingly to changes in time:

$$n_+(t) = \frac{n_{+o}}{1 + n_{+o} k_r t} \tag{8}$$

and

$$n_e(t) = \frac{n_{eo}}{1 + n_{+o} k_r t} e^{-k_a N_a t} \tag{9}$$

With the steady state solutions for n_e (Eq. 4) and for n_+:

$$n_{+o} = \sqrt{\frac{S}{k_r}}, \tag{10}$$

the electron density is given as:

$$n_e(t) = \frac{S}{\left[k_a N_a + \sqrt{k_r S}\right]} \frac{1}{\left[1 + \sqrt{k_r S}\, t\right]} e^{-k_a N_a t}. \tag{11}$$

Another way to get insight into switch operation after ionizing source termination is given by considering the rate of change in electron density (Eq. 1) for either recombination or attachment dominated discharges. For a non-attaching gas ($k_a=0$; $n_-=0$), the relative change of electron density, n_e, (and the switch conductivity, σ) is proportional to the instantaneous value of n_e:

$$\frac{dn_e/dt}{n_e} = -k_r n_e, \tag{12}$$

i.e., the relative rate of change slows down with reduced electron density. In an attachment dominated discharge ($k_a N_a \gg k_r n_{+o}$) on the other hand, the relative rate of decay is <u>not</u> dependent on the electron density:

$$\frac{dn_e/dt}{n_e} = -k_a N_a \tag{13}$$

It, therefore, can be controlled completely by the type and concentration of the electronegative gas added to the buffer gas.

In order to achieve opening times of less than one microsecond at initial electron densities $n_{eo} < 10^{-14}$ cm^{-3}, the dominant loss mechanism must be attachment. That means that the switch gas mixture must contain an electronegative gas. On the other hand, additives of attaching gases increase the power loss during conduction.

Both low forward voltage drop and fast opening can be obtained by considering the E/N dependence of rate coefficients and transport coefficients, and choosing gases or gas mixtures which satisfy the following conditions:[7,9,17,18]

1. For low values of the reduced field strength E/N, characteristic for the conduction phase, the electron drift velocity, w, should be large and the attachment rate coefficient, k_a, should be small in order to minimize losses.

2. With increasing E/N, characteristic for the opening phase of a switch in an inductive energy storage system, the attachment rate coefficient should increase and the electron drift velocity should decrease in order to support the switch opening process.

3. Additionally, the gas should have a high dielectric strength to hold off the expected high voltage across the switch when it opens.

For a given gas mixture, the number of processes which have to be considered in the rate equations may be significantly larger than in Eqs. (1)-(3). Ionization processes in the discharge, especially from excited states, will influence the opening phase at high values of E/N and must be considered to predict the transition into the self-sustained mode which is equivalent to a loss of the external control. Therefore, rate equations for significant excited state populations must be added. At high electron densities, collisional electron detachment must be considered. If additional optical control mechanisms are used, radiative processes also have to be incorporated.[9,16] At high attachment rates during switch opening, the ion currents may also have to be considered.

The major difficulty in solving such a set of rate equations is the necessity to provide the E/N-dependence of all rate constants and mobilities in the E/N range of the discharge during a full switch cycle. In many cases the rate constants have not been measured for the specific gas mixtures and only cross sections of the considered processes are available. In these cases, first a separate Boltzmann or Monte Carlo analysis of the electron energy distribution function, including the influence of the electron beam as a source,[5] is required, with E/N as the variable parameter. The special source type can significantly influence the electron energy distribution function. For an UV sustained discharge, for example, the energy of the initial secondary electrons is low, not significantly changing the steady state distribution function at low E/N. For an e-beam sustained discharge, however, the initial secondary electrons may have significant energies. These electrons will, at low E/N during the relaxation, allow processes at significant rates which otherwise would not be expected at these E/N values.[5,19,20]

DIFFUSE DISCHARGE OPENING SWITCHES

The rate constants are then calculated using the electron energy distribution functions for different values of E/N and the measured cross section.[9] Both, Boltzmann and Monte Carlo calculations, however, require as complete a set of cross sections for the used gas as possible. It is therefore understandable that this approach is possible only for a small number of cases.

Calculations on the discharge behavior of specific gas mixtures have been performed by several authors. Fernsler et al.[14] studied the influence of an admixture of an attacher (O_2) to a buffer gas (N_2) during the opening phase of the discharge. The attachment rate of the three-body attachment process (e+O_2+N_2 → O_2^-+N_2) was considered to be dominant and independent of E/N. Shortening of switch opening times with increasing attacher concentration was demonstrated.

Kline[15] presented calculations on discharges in N_2, Ar, N_2:Ar mixtures, and CH_4 for the operating range of small e-beam current densities (some mA/cm^2). The steady state analysis presents the discharge characteristic, the switch characteristic, and the current gain. The current gain in an e-beam sustained discharge is considered a figure of merit for the control efficiency of the diffuse discharge switch. It is defined as the ratio of switch current to sustaining e-beam current. The e-beam current density J_b is related to the source function S through the following equation:

$$S = \frac{dW}{dx} \frac{J_b}{eW_i}, \tag{14}$$

where dW/dx is the spatial rate of e-beam energy loss in the gas[21,22] and eW_i is the mean ionization energy of the gas. The magnitude of eW_i is in the order of twice the ionization potential of the species.[23].

The switch current density J_s (using Eq. 4) is given as:

$$J_s = \sigma E_s = \frac{e\mu S}{k_a N_a + \sqrt{Sk_r}} E_s, \tag{15}$$

E_s being the electric field intensity in the diffuse plasma. By combining Eqs. (14) and (15), we obtain the expression for the current gain:

$$\frac{I_s}{I_b} = \frac{J_s}{J_b} = \frac{e\mu \, dW/dx}{eW_i(k_a N_a + \sqrt{Sk_r})} E_s \tag{16}$$

For recombination dominated discharges ($k_a = 0$), the current gain scales with $1/\sqrt{S}$, for attachment dominated discharges ($k_a N_a \gg k_r n_{+o}$), the current gain becomes independent of S.

Kline's transient analysis of the diffuse discharge incorporates the influence of an attacher; however, an attachment rate independent of E/N was considered. As a result of the calculations, methane was found to be the best of the switch gases studied since it exhibits a high electron drift velocity at low values of E/N. Fig. 2 shows the measured and predicted e-beam switch waveforms in methane for the experimental conditions of Hunter.[24] The three different predictions result from different values for the recombination and attachment rates used in the calculations.

Schaefer et al.[9,16] presented calculations on discharges in N_2 with N_2O as an attacher. These calculations concentrate on the influence of an attacher with an attachment rate increasing with E/N. It was demonstrated that such an attacher with sufficient concentration generates a discharge characteristic with a negative differential conductivity (NDC) in an intermediate E/N-range. This

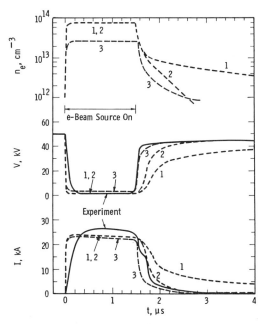

Fig. 2. Measured[2] and predicted[15] e-beam switch waveforms in methane. Impurity attachment rates:
(1) $k_a N_a = 0$;
(2) $k_a N_a = 7.2 \times 10^5$ s^{-1};
(3) $k_a N_a = 2.7 \times 10^6$ s^{-1}.

behavior has also been predicted for discharges in N_2 with admixtures of CO_2 and O_2 as attachers.[25] Figure 3 shows the steady state E/N-J characteristic with the source function S as the variable parameter for an N_2/N_2O discharge.[9] Below ~ 3 Td, the discharge is recombination dominated. Above ~ 3 Td, the attachment rate increases, causing the conductivity to decrease (NDC). This effect is more pronounced at small values of the source function S. At high values of the source function, the electron density will be high enough so that recombination is also significant in the E/N range with a high attachment rate and the negative differential conductivity disappears. Such an NDC characteristic will be found for any attacher with an attachment rate coefficient which has a threshold at some value of E/N and is strongly increasing above this value. A threshold in the range of 1-3 Td seems appropriate, considering that the discharge should be operated at low values of E/N during the conduction phase to achieve low losses. The transient analysis shows that such a discharge characteristic allows operation of the discharge with low losses during the conduction phase in the E/N range below the threshold for attachment and achievement of fast opening times.

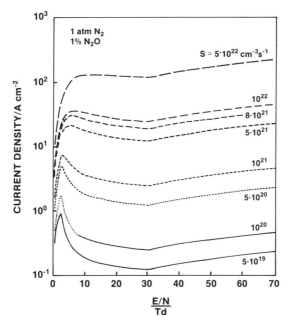

Fig. 3. Calculated steady state E/N-J characteristics for an e-beam sustained discharge in N_2 with an N_2O fraction of 1%. The variable parameter is the electron generation rate.[16]

Figure 4 shows the calculated time dependence of E/N and the current density, J, with the attacher concentration as the variable parameter. For the three smaller values of the attacher concentration (0.1%-0.75%), the discharge reaches the same low value of E/N where it is recombination dominated. The opening time decreases with attacher concentration, however, the closing time increases. For the higher values of the attacher concentration ($\geq$ 1%) the discharge is obstructed from reaching the low E/N range and stays attachment dominated, causing high losses. These results demonstrate that the desired attachment rate for low losses during conduction and for fast opening obstructs switch closing. Photodetachment is therefore discussed as an additional control mechanism to overcome attachment during switch closure (see "Optical Control Mechanisms").

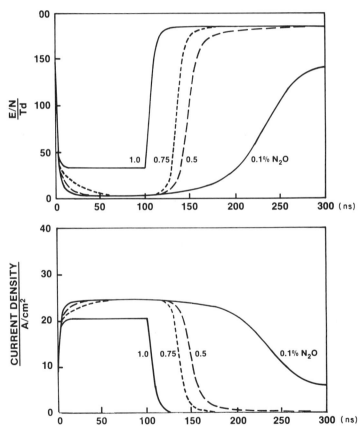

Fig. 4. Time dependence[16] of E/N (top) and current density (bottom) of an e-beam sustained discharge in 1 atm N_2 with admixtures of N_2O. The e-beam is on for $0 \leq t \leq 100$ ns. The variable parameter is the N_2O fraction in %.

BOUNDARY EFFECTS AND INSTABILITIES

The voltage drop across e-beam sustained diffuse discharges ranges from several hundred volts to several kilovolts, depending on the operating conditons. An appreciable amount of the total voltage will be cathode fall voltage. Experimental and theoretical studies of the influence of the cathode fall on the diffuse discharge characteristic have been performed by Lowke and Davies,[12] Bletzinger,[26,27] and Hallada et al.[28] in the United States. Information about research efforts in this field in the USSR can be found in the review of Bychkov et al.[13]

The cathode fall voltage is a function of e-beam current, discharge current, gas type and pressure and electrode material. Figure 5 shows the cathode fall voltage as a function of e-beam current density[29] for discharge current densities of 0.33 and 0.67 A/cm^2. The influence of discharge current is small. Even at large e-beam currents, the cathode fall voltage decreases very slowly. For atmospheric pressure discharges and small gap distance, it can be considerably larger than the voltage across the bulk of the discharge. The addition of an attaching gas does not seem to increase the cathode fall voltage very much. The observed increase in discharge voltage is due mostly to an increase of the voltage drop across the main volume of the discharge.

A theoretical investigation of the cathode fall in high pressure, e-beam controlled discharges was conducted by Hallada et al.[28] Secondary emission from the cathode and electron-impact ionization of metastable states were included in their analysis. The results showed that the cathode fall of an e-beam ionized discharge is very sensitive to the value of the cathode secondary emission coefficient, demonstrating the importance of the electrode properties.

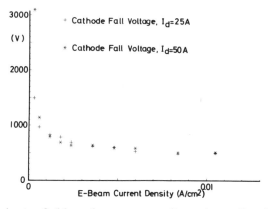

Fig. 5. Cathode fall voltage as a function of e-beam current density (after the foil).[29]

Instabilities which lead to a glow-to-arc transition usually originate near the electrodes, a region of high electric field intensity and lower electron density. Streamer formation is considered to begin with the formation of a filament of gas, heated to a temperature adequate to provide sufficient thermal ionization to become significantly conductive. In addition, streamers can grow from cathode layer hot spots, and have been observed to originate on cathode protrusions.[30] The streamers subsequently propagate at speeds close to sonic, generally progressing with increasing velocity across the gap. An important feature of streamers in discharge gases is, therefore, that arcing can occur a considerable time after the e-beam current has been terminated.[31]

The development of plasma instabilities in the bulk of the diffuse discharge has been addressed by Haas,[32,33] Nighan,[34] and Bychkov et al.[13] Two general models of bulk instabilities were proposed, thermal and ionization. The thermal instability is due to local heating and dilution of the gas with subsequent increase in E/N, which in turn causes increased local energy deposition, etc. The ionization instabilities occur in high field regions of the discharge, preferentially close to the electrodes, when the regenerative ionization becomes comparable to the ionization caused by the external source. The times for the development of these instabilities range from hundreds of nanoseconds to milliseconds, depending on discharge parameters and filling pressure.[35]

Discharges in mixtures containing electronegative gases can exhibit an attachment instability, if the attachment rate coefficient, k_a, is a rapidly increasing function of E/N.[25,36] A small local increase in the electric field intensity in such a gas mixture causes an increase in attachment which leads to a decrease of the electron density and therefore to a further increase in electric field. Ultimately, waves of high electric field in electron-depleted regions move across the discharge, resulting in current waves. The relevance of this instability is that the high local fields caused by the electron loss will greatly increase local heating, and streamers are initiated, growing toward both electrodes.

A problem related to instabilities in diffuse discharges is power loading. Power loading effects in gas diffuse models were addressed by Pitchford.[37] The electrical energy density deposited in the gas is the product of discharge current density and electric field, integrated over time. This energy is deposited in the form of kinetic energy of the gas molecules (gas heating) and in the internal degrees of freedom of the gas molecules. Experimental investigations[31] have led to the conclusion that a uniform discharge can be sustained until the gas temperature has increased to approximately 500 K. After that point the uniform glow goes over to an arc.

There have been attempts to quantify the fraction of the electrical energy input which results in gas heating as a function of E/N for timescales short compared to vibrational relaxation.[38] As might be expected, a major fraction of the energy input to molecular gases is consumed by vibrational excitation. Shirley and Hall[39] determined vibrational temperatures in H_2 as a function of energy density deposited in the gas. They found vibrational temperatures of 1500 K to 1800 K for power loadings of several kJ/mole, or about 0.1 J/cm^3-atm at 30 Td. The changes due to enhanced population in the high energy states of the gas molecules will affect the behavior of the diffuse discharge switch, especially if attachers are used where the attachment cross section changes drastically with vibrational excitation such as in HCl or H_2.[40,41]

OPTICAL CONTROL OF DIFFUSE DISCHARGES

External control of a diffuse gas discharge, in general, infers making use of some mechanisms to influence the charge carrier balance in the discharge, in addition to those processes solely caused by the applied electric field. The control mechanism can be used to increase or to decrease the conductivity by controlling either the electron generation or depletion mechanism.

For opening switches, mainly externally sustained discharges are considered. An externally sustained discharge is operated in the E/N range where the ionization coefficient is negligible so that the electron generation process is given solely by the external source. Besides electron beams, as discussed later in this chapter, UV radiation has been used as a sustainment method for TEA-laser discharges.[42,43] Similar devices with glow discharges sustained by high power flash boards have also been investigated for switching applications.[44-46] The advantage of these devices is that the design and operation of a UV flashboard is significantly simpler than an electron beam gun. For efficient impedance matching, a large number of sparks are operated in series.[44] However, a problem is to achieve a fast cutoff of the ionization source. The propagation time for the exciting electric pulse through the flashboard[46] and its long afterglow[44] seem to limit the opening time to values of some microseconds.

An alternative would be to use lasers which allow a very precise timing. Due to the relatively high costs of laser photons, however, it seems to be inefficient to use lasers as a sustainment method, especially if high photon energies are required. Lasers, on the other hand, offer the unique possibility to influence the density of specific states of one of the species generated in the discharge, and subsequently the generation and depletion mechanisms of electrons as well as insure precise timing as pioneered by Guenther.[8] Laser discharge control for switching application is, therefore, considered now as an additonal control for a specific phase of the switch cycle, where other means do not allow for optimization of the discharge properties.

Laser control of diffuse discharges for opening switches is a very recent research field. Besides papers on concepts,[7,8,47] model calculations on certain proposed systems,[9,16] and investigations of basic processes suitable for switching applications,[48,49] there are no results on operational switching devices available at the present time.

As discussed in the section on "Diffuse Discharge Analysis," the main effort concentrates on optimization of the electron depletion through attachment. Although an optimization of the steady state phase and the opening phase is possible through tailoring the E/N dependence of the attachment rate coefficients, there are still problems related to the closing phase. Concepts for additional optical control, therefore, concentrate on attachment and its control in one of the transition phases. In principle, there are two different methods for optically controlling attachment: (1) to use photodetachment to overcome attachment,[47] or (2) to use optically induced attachment in gases which otherwise would not have a strong attachment rate in the E/N range of interest.[7]

Photodetachment

Photodetachment can be used as a method to overcome attachment in a well defined discharge period. Just like photoionization, photodetachment is a nonresonant process, which in general requires much lower photon energies. The cross sections usually are not very high, requiring a high laser power. Therefore, photodetachment seems to be an appropriate process only if needed for a short switch period in discharges sustained by other means.[9]

A negative ion considered for photodetachment[47] as a control mechanism is O^-. The photodetachment cross section (see Fig. 6) has a threshold at a photon energy of 1.47 eV and reaches a plateau of 6.3×10^{-18} cm^2 at approximately 2 eV.[50] Several molecules, such as O_2, NO, CO, N_2O, NO_2, CO_2, and SO_2, undergo dissociative attachment producing O^-, e.g.,

$$e^- + O_2 \rightarrow O^- + O.$$

In fact, in all these gases or gas mixtures with these gases, competitive attachment processes exist. Frequently, subsequent reactions of O^- will occur, producing molecular negative ions which in general have lower cross sections for photodetachment. In discharges in O_2, for example, the dominant molecular negative ion is O_2^- which is produced either through attachment in a three-body collision:

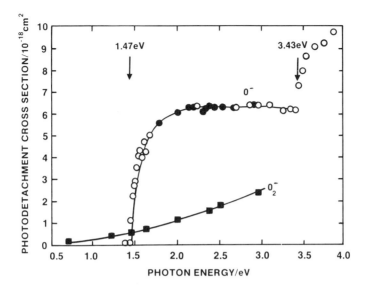

Fig. 6. Photodetachment cross section for O^- and O_2^- [50,52,53].
- ● Smith
- o Branscomb et al.
- ■ Burch et al.

$$e^- + O_2 + O_2 \rightarrow O_2^- + O_2,$$

which is the dominant attachment process at high pressures,[51] or in a charge transfer collision with O^-:

$$O^- + O_2 \rightarrow O_2^- + O.$$

The cross sections for photodetachment of O^- and O_2^- are shown in Fig. 6,[50,52,53] and are different by a factor of approximately 5 at 2 eV. Photodetachment experiments[47] in low pressure O_2 discharges and flowing afterglows indicate that a fraction of 50% of the negative ions can be detached with laser pulses with an energy flux of 35 mJ/cm^2 and a photon energy of 2.2 eV.

For three-atomic molecules, either a dominant branch for dissociative attachment producing O^- exists, as in N_2O:

$$e^- + N_2O \rightarrow O^- + N_2,$$

or several branches producing O^- <u>and</u> molecular ions, as in SO_2 with similar cross sections:[54]

(1) $e^- + SO_2 \rightarrow O^- + SO$

(2) $e^- + SO_2 \rightarrow SO^- + O.$

It should be mentioned that photodetachment will produce large effects only if the lifetime of the negative ions is long enough (some 10 ns) and no competitive detachment processes will nullify the attachment reaction. For systems producing O^-, for example, it is known that the reactions:

$$O^- + H_2 \rightarrow H_2O + e^-$$

$$O^- + CO \rightarrow CO_2 + e^-$$

can significantly change the negative ion production.[55,56]

For attachers with a high attachment rate at high values of E/N and low or zero attachment rate values of E/N, photodetachment may be a suitable process to support a transition from a highly attaching state into a non-attaching state of an externally sustained discharge. The effect of laser photodetachment on the closing phase of an electron-beam sustained discharge in N_2 containing an admixture of N_2O was calculated by Schaefer et al.[9] Figure 7 shows the transient behavior of the discharge for an operating condition in which the closing phase is obstructed by attachment so that the discharge never reaches the low E/N-regime where it is not dominated by attachment. With a laser power density of 10^7 W/cm^2, attachment is partially compensated and the discharge behaves similarly to one with a lower attachment concentration (compare with Figure 4). Since the laser is required only during the closing phase, for about 10 ns, the power density of 10^7 W/cm^2 corresponds to an energy flux of 100 mJ/cm^2. Since only a small fraction of the power is absorbed, this requirement can easily be fulfilled by a multipass or intracavity optical design.

Optically Enhanced Attachment

Optically enhanced attachment infers the use of a gas mixture with an additive of molecules, which in their intial state are very weak attachers, and the transfer from these molecules, through optical excitation and maybe some subsequent spontaneous transitions, into species which act as strong attachers. Optically enhanced attachment is a control mechanism considered for controlling the opening phase of diffuse discharge opening switches.[7]

Certain attachers have a drastically increased attachment cross section in their rotational and/or vibrational excited states. This mechanism can be understood by considering the potential energy curves of a diatomic molecular attacher and its negative ion. (For a review on attachment see Ref. 57). In Fig. 8, a general type of dissociative electron attachment process is illustrated. The potential energy curve of a neutral diatomic molecule, AB, is crossed at an energy, E_v, above the ground state by a repul-

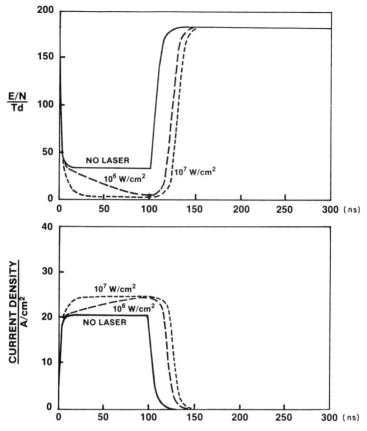

Fig. 7. Time dependence of E/N (top) and current density (bottom) of an e-beam sustained, laser photodetachment assisted discharge in 1 atm N_2 with an N_2O fraction of 1%. The e-beam and laser are on for $0 \leq t \leq 100$ ns. The variable parameter is the laser power density.[16]

sive branch of the negative ion AB^-. In most attachers, several branches of the negative ion AB^- may cross the potential energy curve of the neutral molecule, and the potential energy, $E(A + B^-)$, after dissociation may be below the energy of the molecular ground state, $E(AB)$, as shown in Fig. 8 (e.g., halogens), or above (e.g., oxygen). <u>Dissociative</u> attachment will occur only if the initial energy after electron capture, $E*(AB^-)$, on the branch AB^- is larger than the energy of the dissociated negative ion, $E(A+B^-)$. For a molecule in the vibrational ground state, the electron must provide the energy, $E*$, to allow attachment and succeeding dissociation. This energy determines the maximum of the energy dependence of the cross section for dissociative attachment. For a vibrationally excited molecule, the cross section will depend strongly on the

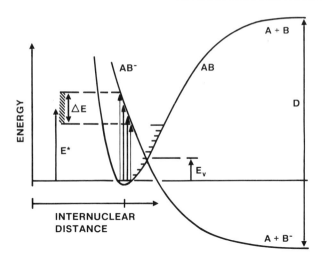

Fig. 8. Resonance dissociative electron attachment.

excitation energy relative to the energy of curve crossing, E_v. For the example shown in Fig. 9, the attachment cross section increases with vibrational excitation up to v = 4, and the electron energy necessary to form the negative ion, AB⁻, and subsequently the threshold of the attachment cross section will shift to smaller values if the molecule, AB, is excited into a vibrational state closer to the curve crossing.

There are two kinds of measurement of the attachment rate indicating whether or not a molecule shows the intended behavior: (1) the E/N dependence, and (2) the temperature dependence of the attachment rate. For a system, as shown in Fig. 8, dissociative attachment will occur only if the electrons provide the energy, E^*, which indicates that the cross section has its maximum at this value and a width, ΔE.[58] Thus, the attachment rate, $k_a(E/N)$, will strongly increase with E/N as long as the mean electron energy, E_e, is below the value E^*, and will decrease for $E_e > E^*$. An example[59] (NF_3) is shown in Fig. 9. For other examples see Gallagher et al.[60] and Christophorou.[57] Therefore, the value of E/N, for which the attachment rate reaches its maximum, is a measure for the value E^* in Fig. 8. The temperature dependence of the attachment rate constant, $k_a(T)$, is a measure of the change of the cross section with vibrational and rotational excitation since the density of excited molecules increases with T. There are several attachers showing a strong increase of the attachment rate with increasing temperature.[58,61]

As yet, there are no known measurements of attachment cross sections for individual vibrational states. In experiments by

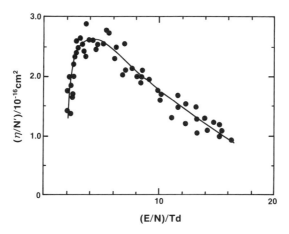

Fig. 9. Effective attachment rate η/N' in 0.5% NF_3 in 20 Torr He. N' refers to the NF_3 gas density.[59]

Srivastava and Orient,[54] however it was demonstrated that monoenergetic electrons at least allow the excitation of specific modes of vibration, and subsequently the measurement of attachment cross sections of a special group of vibrational states. Bardsley and Wadehra[40] have calculated the cross section for the vibrational states (v=0-3) of HCl (Fig. 10) from cross section measurements at different gas temperatures by Allan and Wong.[61] The population density of the rotational states was assumed to be a Boltzmann distribution at T = 300 K. These data show that the attachment rate can be increased dramatically using vibrational excitation, but this advantage disappears in HCl if the electron energy is larger than 0.5 eV. For switching application, one should look for molecules having their maximum attachment cross section in the ground state at a somewhat higher energy, depending on the E/N-range considered for the conduction phase of the switch.

Calculations of the dissociative attachment cross section for individual vibrational and rotational states (v,J) have been performed by Wadehra,[41] demonstrating that vibrational excitation is more efficient than rotational excitation for increasing the attachment cross section.

There are many ways to optically populate vibrational states. The first experiment performed using laser excitation to increase the attachment rate was done in SF_6 by Chen and Chantry[62] for <u>dissociative</u> attachment of SF_6:

$e^- + SF_6 \rightarrow SF_5^- + F.$

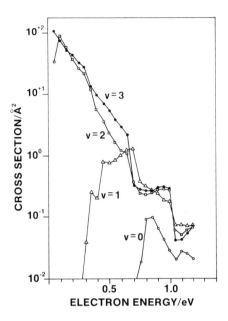

Fig. 10. Theoretical cross section for e + HCl → H+Cl⁻, depending on electron energy for the vibrational states (v=0-3), assuming a Boltzmann distribution over rotational states at T=300 K.[40]

Their experiments showed a significant increase of the attachment rate for this dissociative branch. The fact that this experiment was S-isotope specific, as seen from the laser wavelength dependence, indicated that the absorbed laser power does not result simply in a general gas heating. Recently, similar experiments have been performed by Barbe et al.[63] using the valence electrons of Rydberg Ar-atoms instead of free electrons. Of course, SF_6 is not a suitable molecule for externally controlled attachment, since it is already a strong attacher in the ground state due to the non-dissociative process:

$$e^- + SF_6 \rightarrow SF_6^- (v,J).$$

For application in diffuse discharges, one must consider that vibrational excitation also will result from electron collisions. As the discharge is considered to be cold, mainly collisions with molecules in the ground state occur. This is true for electrons as well as for excited molecules, which makes the excited state with the lowest energy the only excited species at considerable densities. Thus, an external control mechanism will work efficiently only if a significant increase in the attachment cross section of the considered electron energy range occurs for states v > 1 for two atomic molecules, or for large molecules for states with energies well above that of the lowest lying excited state. Some possible mechanisms for producing molecules in vibrationally excited states which are not strongly created in a cold discharge are listed in Table 1. Examples of most of these processes can be found in excitation mechanisms for lasers working between vibrational states.

Table 1. Photoenhanced electron attachment

---

(1) VIBRATIONAL EXCITATION

  (a) Single-Photon Excitation plus Energy Transfer
  (b) Overtone and Combination Band Excitation
  (c) Multi-Photon Excitation

---

(2) PHOTODISSOCIATION

  (a) Single-Photon - UV
  (b) Multi-Photon - IR

---

(3) ELECTRONIC EXCITATION

  (a) $E \rightarrow v,J$ Collisions
  (b) $E \rightarrow v,J$ Transitions, Radiative

---

(1) Optical vibrational excitation of higher lying states can be accomplished through vibration-vibration energy transfer, through overtone excitation or excitation of combination states, and through multi-photon excitation. Such mechanisms can be found in numerous excitation mechanisms of optically pumped IR lasers and therefore will not be discussed here in detail. Some examples are given in a paper by Tiee and Wittig,[64] in which laser action is reported in molecules such as CF_4, $NOCl$, CF_3I, and NH_3 by optically pumping with a CO_2 laser. To date, however, experiments have failed to demonstrate a significant increase of the attachment rate in an electron beam sustained discharge after irradiation with a pulsed CO_2 laser.[49]

(2) Single- and multi-photon dissociation of large molecules also has been shown to produce molecules in vibrationally excited states and also is used as an excitation mechanism for molecular gas lasers. One example of each process is presented here. Sirkin

and Pimentel[65] demonstrated that photoelimination of HF from CH_2CF_2 and CH_2CHF, using a UV-flashlamp, would generate a significant fraction of the HF molecules in states v > 1. Other molecules generating HF molecules in vibrationally excited states through UV photoelimination are $CH_3CH_{3-x}F_x$ with x = 1,2,3 and similar molecules with more than 2 C-atoms. With chlorinated hydrocarbons, equivalent HCl molecules are produced.[66] Some of these processes have significant cross sections at the wavelength of the ArF-laser at 193 nm. Rossi et al.[48] performed drift tube experiments to demonstrate the feasibility of these processes for controlling the electron balance in a discharge. Figure 11 shows their experimental results. In a 100 torr mixture of helium with 100 mtorr C_2HF_3 at low values of E/N (< 3 Td), they obtained an increase of the attachment rate of up to 10^3 with an ArF laser at 193 nm. Similar experiments were performed in C_2H_3Cl. Schaefer et al.[9] performed measurements in a DC discharge at low values of E/N (1-10 Td), which was externally sustained by helium plasma injection. In a gas mixture of 60 torr helium and 3% C_2H_3Cl, pulsed resistance

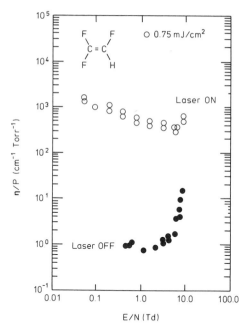

Fig. 11. Attachment coefficient for trifluoroethylene. The solid dots give the values for the unexcited sample (200 mTorr of C_2F_3H in 200 Torr helium). The open circles represent the data for the excited sample (100 mTorr of C_2F_3H in 100 Torr helium). The attachment coefficients are expressed in terms of the unexcited trifluorethylene pressure.[48]

changes of a factor of 3.5 were measured after irradiating the discharge with a UV spark source through a quartz window.

Vibrationally excited molecules can also be generated through multi-photon processes with IR lasers. Quick and Wittig[67] used a CO_2 laser for multi-photon dissociation of several fluorinated ethanes and ethylenes to produce vibrationally and rotationally excited HF molecules with a significant fraction in $v > 1$.

(3) Electronic excitation can be transferred into vibrational excitation either through collisions or radiative processes (for a review see Ref. 68). Collisional transfer subsequent to optical excitation has, for example, been used to operate a CO_2 laser. Petersen, et al[69] used the following processes:

$$Br_2 + h\nu \rightarrow Br + Br^*$$

$$Br^* + CO_2 \rightarrow Br + CO_2^\dagger + \Delta E$$

where the dominant excited states of CO_2 are the $(10^0 1)$- and $(02^0 1)$- states.

Electronic excitation and subsequent radiative transitions into highly vibrationally excited states have been used by Beterov and Fatayev[70] to show that vibrational excitation of I_2 can increase the attachment rate for dissociative attachment. Figure 12 depicts the excitation mechanism. The radiation from a frequency doubled Nd:YAG laser ($\lambda = 532$ nm) was used to excite the (B)-state. Intense Stokes flourescence indicated a subsequent transition into highly vibrational excited states of the electronic ground state. Vibrational relaxation had to provide the population of the optimum vibrational states, resulting in an increase of the attachment rate by three or four orders of magnitude. Since the vibrational relaxation required times in the order of some 10 μs, it also was suggested to use a stimulated resonant Raman process to excite directly the vibrational state with the highest attachment cross section.

For all proposed processes of optically enhanced attachment, it is very important to consider the attachment properties of the starting compound. A real advantage of optically enhanced attachment can be achieved only if the attachment rate is significantly increased over a wide E/N range. If the starting molecule has an attachment rate coefficient strongly increasing with E/N, it also will generate a negative differential conductivity, obstructing the closing process. Also, the stability of the starting molecule with respect to collisional processes in the e-beam sustained discharge must be investigated. At this time, UV photodissociation of molecules generating attaching fragments looks most promising. The best starting molecule, however, may not yet have been found.

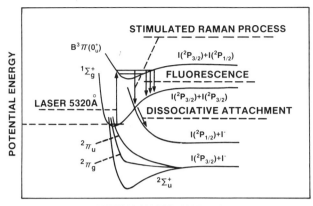

Fig. 12. Some potential curves for I_2 and $\bar{I_2}$ and the scheme of excitation of vibrational levels.[70]

ELECTRON-BEAM CONTROL OF DIFFUSE DISCHARGES

Electron Beam Considerations

Electron beams are the most commonly used ionization sources for externally sustained diffuse discharges. Typically, beam energies are in the 100-300 keV range with current densities up to several A/cm^2. In order to generate the e-beam, it is necessary to have the electrodes in a vacuum of $< 10^{-3}$ Torr for cold cathode operation and $< 10^{-6}$ Torr for thermionic cathodes.

The electron generation mechanism at cold cathodes is usually field emission. With field-emitting cathodes, current densities of several hundred A/cm^2 can be obtained[71] at the expense, however, of limited pulse duration due to "diode closure".[72] Diode closure is caused by plasma formation at the cathode. The plasma moves with velocities of 2-5 x 10^6 cm/s toward the anode and "closes" the diode gap in typically 2-5 µs (or, in some very high current density situations, much faster).

A type of field emission cathode fabricated using film techniques and microlithography has been investigated by Spindt et al.[73] The emission originates from the tip of molybdenum cones that are about 1.5 µm high with a tip radius of around 500 Å. Maximum currents in the range of 50-150 µA per cone can be drawn with voltages in the range of 100-300 kV when operated at pressures of $< 10^{-9}$ Torr. In arrays, current densities of > 10 A/cm^2 have been demonstrated. The film field emission cathode allows separation of the circuit for electron emission, where voltages of about 100 V are sufficient to control the electron flow, and the acceleration circuit for the electrons. So far, however, fabrication problems seem to limit the application of this type of cathode.

Another cold cathode used for electron generation in e-beam controlled diffuse discharges is the wire-ion-plasma (WIP) electron gun.[74,75] An experimental arrangement[76] is shown in Fig. 13. The WIP gun cavity is usually filled with helium, typically at 10-20 mTorr. A positive pulse applied to the wire anode ionizes the helium in the plasma chamber. Helium ions are extracted and accelerated by the DC high voltage applied to the e-beam cathode. Upon helium impact, electrons are emitted from the cathode. The electrons are accelerated by the high voltage, V_{eb}, and transmitted through the grids. The e-beam current obtained in this device is typically 10 A over a cross-section of 100 cm^2.

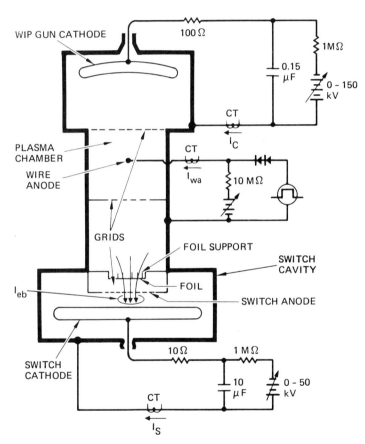

Fig. 13. Schematic of an electron beam controlled switch and its test circuit. The controlling electron beam is obtained from a Wire-Ion-Plasma Electron Gun.[76]

Thermionic emitters offer the advantage of decoupling electron generation and electron acceleration with (in principle) unlimited pulse length. The generation of pulse trains is possible by using a triode or tetrode configuation. In a tetrode with thoriated tungsten electrodes,[77] current densities of up to 4 A/cm^2 were obtained at a cathode temperature of 2000° C. With dispenser type thermionic cathodes, values of 300 A/cm^2 were reached.[78] (For a review on dispenser cathodes see Ref. 79). For a dispenser cathode, a material with a low work function, e.g., barium, is integrated into a porous tungsten matrix. Heating dispenses barium from the interior to the surface, thereby continuously replenishing the surface with low work function material.

The interface between the electron gun and the switch chamber is generally a thin (20 to 50 μm) metal foil backed by a support structure. It should have low electron stopping power, and good thermal and mechanical properties. In Table 2, characteristics of different foil materials are listed[71]. A severe constraint for long e-beam pulses and/or repetitive operation is foil heating. In order to avoid structural fractures, the foil temperature must be limited. A foil heating analysis has been performed by Daugherty.[80]

Design of an E-Beam Controlled Diffuse Discharge Switch

Because of the limited conduction time of an e-beam sustained discharge due to the development of instabilities and the time constraints of the e-beam source, it is proposed to use it in a two-loop inductive energy storage circuit.[9] Figure 14 shows the schematic circuit with the charging loop and the discharge loop having only an inductor in common. The inductor is charged through the initially closed slow opening switch. The e-beam is turned on when the current through the inductor has reached its maximum steady state value. The slow opening switch can now open the charging circuit at relatively low switch voltage and commutate the current into the e-beam controlled discharge. If switch gases with low losses at low E/N are used, the efficiency of this commutation can be very high. Once the current is transferred into the diffuse discharge switch, the circuit can generate either a single pulse by closing the closing switch and simultaneously opening the diffuse discharge, or a pulse train by repetitively closing and opening the diffuse discharge with the closing switch closed. In order to prevent loss currents in the load when the diffuse discharge is conducting (if the load impedance is comparable to the diffuse discharge impedance), the closing switch could be replaced by a second e-beam controlled switch which opens when the first switch closes and vice versa.

Table 2. Characteristics of foil material candidates[71].

FOIL MATERIAL	ADVANTAGE	DISADVANTAGE
Beryllium	Very Low Z High Conductivity	Highly Toxic Available Only in Small Pieces, Expensive ~ $50/in^2
Kapton	Low Z	Very Low Conductivity Disintegrates from Electron Radiation
Aluminum	Relatively Low Z High Conductivity	Low Strength at T > 200°C
Titanium (Pure and Alloy)	High Strength	Low Conductivity Relatively High Z
Stainless Steel	High Strength	Low Conductivity High Z
Inconel 718	Very High Strength	Low Conductivity High Z
Composite e.g. Cu or Al Clad Ti	Combines High Strength with High Conductivity	Difficult to Fabricate in Uniform Layers with Good Bonding

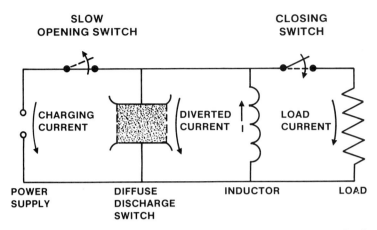

Fig. 14. Schematic circuit for an e-beam sustained discharge switch in an inductive energy storage system.[9]

The following design criteria for the e-beam sustained switch in such a circuit are based on a group report[81] in the workshop proceedings on "Diffuse Discharge Opening Switches". For an efficient switch, the major electron loss in the conduction state will occur through recombination. Attachment is important only in onset and cutoff, resulting in a steepening of the current pulse. In steady state the electron density is well described by Eq. (4) with k_a being neglected. The conductivity therefore is given by Eq. (7) with Eq. (14):

$$\sigma = e\mu \sqrt{\frac{\frac{dW}{dx} J_B}{eW_i k_r}} \quad \text{with } \mu = \frac{v_d}{E_s}, \qquad (17)$$

where v_d is the electron drift velocity.

In order to avoid glow-to-arc transition, the temperature in a molecular gas should not exceed $T_o \cong 500$ K. Assuming that any increase in gas temperature, ΔT, is caused by Joule heating, leads to the relation:

$$\rho c_v \Delta T = \frac{J^2 \tau}{\sigma}, \qquad (18)$$

ρ being the gas density, c_v the specific heat, and τ the conduction time. This relation allows estimation of an upper limit for the current density:

$$J_{max} = \sqrt{(T_o - T_i)\frac{\rho c_v \sigma}{\tau}}, \quad (19)$$

where T_i is the initial temperature.

For a given total current, I, this determines the switch diameter. The discharge gap length d is determined by the required switch holdoff voltage, V_o:

$$d = \frac{V_o}{E_{crit}}. \quad (20)$$

Critical field strengths, E_{crit}, are in the order of 6-12 kV/cm at pressures of ~ 1 atm.

Calculations along these lines were performed[81] for a switch which carried 10 kA for a time of 50 μs to charge a capacitor of 2.0 μF to a voltage of 250 kV. A switch gas pressure of 10 atm in 90% CH_4 and 10% Ar was assumed. The standoff capability of this mixture is E_{crit} = 5 kV/cm at 1 atm. For a holdoff voltage of 250 kV, the gap length, d, must be at least 5 cm at a pressure of 10 atm. With an assumed e-beam current density of J_b = 140 mA/cm^2 and an e-beam energy of 200 keV after passing the foil which separates e-beam gun and diffuse discharge switch, the discharge conductivity is $\sigma = 2.1 \cdot 10^{-2}$ (Ω cm)$^{-1}$ (Eq. (17)). The maximum current density is, according to Eq. (19), $J_{max} \approx 19$ A/cm^2. For a total current of 10 kA, the switch area must be ~ 530 cm^2.

Using these data, the transfer efficiency of the switch can be calculated. The energy stored in the capacitor is E_{cap} = 62.5 kJ. The Joule losses in the diffuse discharge are E_{dis} = 2.25 kJ. The e-beam energy dissipated in the switch is 0.75 kJ. Allowing, conservatively, structure and foil e-beam losses to be 30%, the total required e-beam energy is E_{eb} = 1.06 kJ. The efficiency is then given as

$$\eta = \frac{E_{cap}}{E_{cap} + E_{eb} + E_{dis}}, \quad (21)$$

which is about 95% for this diffuse discharge switch. This calculation allows a rough estimate of the geometry and the transfer efficiency of diffuse discharge switch as part of a pulsed power circuit.

A more detailed design procedure which includes the energy consumption during opening of a repetitively operated switch was developed by Commisso.[82] The energy transfer efficiency, η, is defined as:

$$\eta = \frac{I_L^o V_L^o \tau_L}{\langle I_{sw} V_{sw}\rangle \tau_c + \langle I_{sw} V_{sw}\rangle \tau_o + I_L^o V_L^o \tau_L + I_{eb} V_{eb} \tau_c} , \qquad (22)$$

where I_{sw}, I_L and I_{eb} are the switch, load and e-beam currents, V_{sw}, V_L, and V_{eb} are the corresponding voltages (the superscript o denotes peak values). τ_c, τ_o. and τ_L are the conduction time, the opening time, and the characteristic load pulse width, respectively. The terms in the denominator describe Joule losses in the switch during conduction and during opening, the energy transferred to the load, and the e-beam energy necessary to sustain the discharge for the time of conduction.

For an energy transfer efficiency comparable to capacitive store pulsed power generators, it is required that each of the losses in the switch is small compared to the energy transferred to the load. These conditions can be interpreted as restrictions on the switch opening time τ_o, on the voltage drop across the discharge during conduction V_{sw}, and on the current gain $\epsilon = I_{sw}/I_{eb}$.

The system was optimized by requiring that the switch pressure be minimum. The result of this optimization is that each of the major energy losses--conduction, opening, and electron-beam production--are roughly equal to each other.

The transfer efficiency calculated by using this formalism,[82] which includes switch losses during opening, is smaller that the one obtained with the simple model discussed before Eq. (21). The design equations for the switch geometry in the two models become identical for the case that the switch losses during opening can be neglected.

Experimental Results

A summary of relevant experiments on e-beam controlled switches is given in Table 3. Listed are e-beam and switch parameters. The values for current gain and opening time are mostly approximate values which serve to define the range of operation for the respective experiment.

Early experiments on diffuse discharge switches were performed by Hunter[2] and Koval'chuk and Mesyats.[3,4] They used cold cathode e-beam emitters derived from laser experiments. With their large e-beam current density, they achieved fast rise times and could afford to work with added attachers to get fast opening, at the expense of power gain. They did not use specially designed gases, however. Hunter recognized the importance of having a gas with a high electron drift velocity and therefore chose methane. Methane has an electron drift velocity peaking at more than 10^7 cm/s at about 3 Td.

Experimental studies in the Soviet Union on diffuse discharge opening switches were initiated at the Atmospheric Optics Institute[3,4] and continued at the Institute of High Current Electronics, both in Tomsk. After a few publications,[83] which were based on the initial studies by Koval'chuk and Mesyats, no further work on this subject appeared until 1982.[84] In this later paper, for the first time, there seems to be more emphasis on optimizing gas properties with respect to their application in switches for inductive energy storage systems, although in former publications the importance of gases with high drift velocity at low E/N was already recognized.

In the United States, several gas mixtures have been proposed for diffuse discharge opening switches.[7,18,24,85] Figure 15 shows electron attachment rate constants for several attachers in Ar and CH_4 as buffer gas.[24] These gas mixtures also exhibit a negative differential drift velocity region over a wide range of fractional concentrations of the attaching gas in the buffer gas, as shown in Fig. 16 for C_2F_6 in Ar and CH_4, respectively. Consequently, the current density-reduced field strength (J-E/N) characteristics of e-beam sustained discharges in gas combinations of this type exhibit a pronounced negative differential conductivity range.[86]

Another gas combination which was proposed as switch gas for e-beam sustained diffuse discharge opening switches is N_2O in N_2 as buffer gas.[7] The gas mixture exhibits an E/N dependent electron decay rate which increases by more than a factor of 20 in the E/N range from 3-15 Td.[87] Pure N_2 has an electron drift velocity which does not decrease with E/N and is therefore not the optimum buffer gas. However, for gas mixtures which show a strong increase of the attachment rate, the drift velocity condition is generally of minor importance.

Work done at AF Wright Aeronautical Laboratory,[26,27,29] at Hughes Research Laboratories[76] and at Westinghouse R&D Center[88] has concentrated on high gain systems with opening times in the microsecond range. In the work of Bletzinger,[26,29] special consideration was given to the cathode region and its influence on the discharge characteristic. Work at the Naval Research Laboratory[89-92] and at Texas Tech University[93-95] dealt with repetitive or burst mode opening switch operation. Both groups concentrated on switches with opening times in the submicrosecond range, but tried to achieve reasonable current gains (> 100) by using gas mixtures whose attachment rates and drift velocities vary with applied

Table 3. Summary of data on relevant experiments on e-beam controlled switches

INSTITUTE, REFERENCES	E-BEAM SYSTEM	E-BEAM CURRENT DENSITY $J_B[A/cm^2]$	E-BEAM ENERGY $E_B[keV]$
Inst. High Current Elect. Tomsk. USSR (Koval'chuk and Mesyats[3,4]):	Cold Cathode	1.5	330
Efremov and Koval'chuk[83] Koval'chuk[84]	Thermionic Cathode, Tetrode	0.014	135
Maxwell Labs San Diego, CA (Hunter,[2])	Cold Cathode	2-5	250
AFWAL Wright Patterson AFB (Bletzinger,[27,29])	Hot Matrix Cathode	0.02	175
Westinghouse R&D Center Pittsburgh, PA (Lowry et al[88])	WIP-gun	<0.1	150
Hughes Research Laboratories, Malibu, CA (Gallagher and Harvey[76])	WIP-gun	<0.035	150
Naval Research Lab, Washington, DC (Commisso et al.[89],)	Cold Cathode 200 ns and 1 µs pulse durations	0.2-5	200
Texas Tech University Lubbock, TX (Schoenbach et al.,[94,95] Schaefer et al.[96])	Thermionic Cathode, Tetrode, pulse train	0.1-1.0	250

DIFFUSE DISCHARGE OPENING SWITCHES

GASES	SWITCH CURRENT DENSITY J_S [A/cm^2]	CURRENT GAIN	OPENING TIME
$CO_2:N_2:He$	15	10	200 ns
Natural Gas: $CH_4:C_2H_6:C_3H_8:$ $O_2:CO_2$:other hydrocarbons	30	1000	3 μs-5 μs
CH_4	27	5	400 ns
N_2,Ar,CH_4 with C_2F_6,C_3F_8,NF_3,	2	100-1000	10 μs-1 μs
N_2CH_4	1.9	380	>1 μs (with inductor in switch loop)
CH_4	20	<1000	>1 μs (with inductor in switch loop)
CH_4,N_2 $N_2:O_2$ $Ar:O_2$ $CH_4:C_2F_6$	<50	10-200	50 ns-100 ns
$N_2:N_2O$ $N_2:SO_2$ $N_2:CO_2$ $Ar:C_2F_6$	<15	~15	50 ns

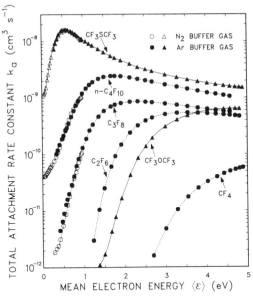

Fig. 15. Total electron attachment rate constants as a function of the mean electron energy $\langle\epsilon\rangle$ for several perfluoroalkanes and perfluoroethers measured in buffer gases on N_2 and Ar.[24]

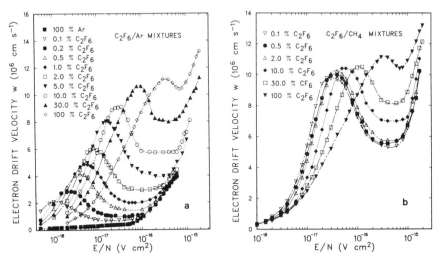

Fig. 16. Electron drift velocity w versus E/N for several (a) C_2F_6/Ar and (b) C_2F_6/CH_4 gas mixtures.[24]

electric field. At Texas Tech University, the interaction of the discharge and circuit has been investigated for discharges with a strong negative differential conductivity, such as in mixtures of Ar and C_2F_6. Figure 17 shows such a characteristic and demonstrates that the maximum current cannot be utilized in the burst mode in a high impedance system if the pulse separation is shorter than the discharge time of the inductor.[96]

Opening switch experiments in discharge circuits including inductors have demonstrated the use of electron beam controlled diffuse discharges for the generation of high voltage pulses. For an experiment performed at Wright-Patterson AFB,[88,89] a capacitor discharge current of up to 135 A was commutated into an inductive load after opening of the e-beam controlled switch. A short (μs) voltage pulse was generated, with its peak voltage exceeding the static breakdown voltage of the e-beam controlled discharge by more than 50% with a CH_4/C_2F_6 gas mixture. An experiment performed at NRL[97] used an e-beam controlled diffuse discharge as the switch element in an inductive energy storage circuit. The storage inductor (1.5 μH) was energized by a capacitor charged to 26 kV. Upon termination of the ionizing e-beam, the 10 kA discharge was interrupted, thereby generating a 280 kV, 60 ns voltage pulse across an open circuit (Fig. 18). The gas mixture in the e-beam controlled switch was CH_4/C_2F_6 at 5 atm. The current gain was about 10.

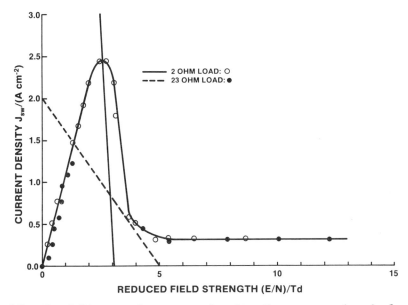

Fig. 17. Load lines and current density J versus reduced electric field strength E/N obtained with different load lines for an e-beam sustained discharge in Ar with an admixture of 2% C_2F_6.

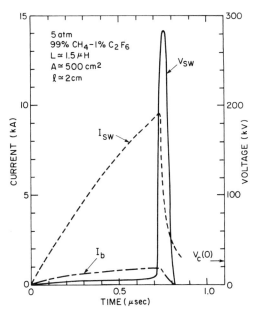

Fig. 18. System current, e-beam current, and switch voltage for an e-beam controlled switch using 5 atm of CH_4 with 1% C_2F_6.[97]

The experimental results demonstrate that diffuse discharges are very suitable for the generation of high power pulses. The low inductance electron beam controlled switch can carry large currents during the charging phase of the inductive storage. When the e-beam is turned off, the switch can rapidly and completely stop conducting in the presence of a high electric field. It can repeat the opening and closing process on a time scale of tens of nanoseconds to microseconds for a burst of pulses.

In order to improve the efficiency (low losses during conduction at high current gain) of these switches and to minimize the opening time, it is important to utilize electron-molecule interactions in the gas mixture more than it has been done so far. Using "Gas Engineering" as a means to design switch gases could lead to diffuse discharge switches with current gains of more than one hundred at opening times of less than 100 nanoseconds. Another very promising approach to increase the efficiency and speed of diffuse discharge switches is the use of optical control mechanisms in electron-beam sustained discharges. "There is no question but that the potential of optically controlled discharges is only beginning to be realized, primarily as a result of improvements in laser versatility and reliability, efficiency of interaction of the optical control energy (particularly in a resonant mode), demonstrated precise timing and synchronization capability to below a picosecond and improved understanding of the total switching event".[98]

REFERENCES

1. J.D. Daugherty, E.R. Pugh, and D.H. Douglas-Hamilton, A Stable, Scaleable High Pressure Gas Discharge as Applied to the CO_2 Laser, Bull. Amer. Phys. Soc., 17:399 (1972).
2. R.O. Hunter, Electron Beam Controlled Switching, Proc. 1st IEEE Inter. Pulsed Power Conf., Lubbock, TX, IC8-1 (1976).
3. B.M. Koval'chuk, and G.A. Mesyats, Rapid Cutoff of a High Current in an Electron-Beam-Excited Discharge, Sov. Tech. Phys. Lett., 2:252 (1976).
4. B.M. Koval'chuk, and G.A. Mesyats, Current Breaker with Space Discharge Controlled by Electron Beam, Proc. 1st IEEE Intern. Pulsed Power Conf., Lubbock, TX, IC7-1 (1976).
5. D.R. Suhre, and J.T. Verdeyen, Energy Distributions in Electron-Beam Produced Nitrogen Plasmas, J. Appl. Phys., 47:4484 (1976).
6. K. Nakanishi, L.G. Christophorou, J.G. Carter, and S.R. Hunter, Penning Ionization Ternary Gas Mixtures for Diffuse Discharge Switching Applications, J. Appl. Phys., 58:633 (1985).
7. K.H. Schoenbach, G. Schaefer, M. Kristiansen, L.L. Hatfield, and A.H. Guenther, Concepts for Optical Control of Diffuse Opening Switches, IEEE Trans. Plasma Sci., PS-10:246 (1982).
8. A.H. Guenther, Optically Controlled Discharges, Proc. ARO Workshop on Repetitive Opening Switches, Tamarron, CO, 48 (1981) DTIC No. AD-A110770.
9. G. Schaefer, K.H. Schoenbach, H. Krompholz, M. Kristiansen, and A.H. Guenther, The Use of Attachers in Electron-Beam Sustained Discharge Switches-Theoretical Considerations, Lasers and Particle Beams, 2:273 (1984).
10. W.H. Long, Modeling of Diffuse Discharge Opening Switches--Group Report, Proc. ARO Workshop on Diffuse Discharge Opening Switches, Tamarron, CO (1982). DTIC No. AD-A115883.
11. W.L. Nighan, and W.J. Wiegand, Influence of Negative-Ion Processes on Steady State Properties and Striations in Molecular Gas Discharges, Phys. Rev. A, 10:922 (1974).
12. J.J. Lowke, and D.K. Davies, Properties of Electric Discharges Sustained by a Uniform Source of Ionization, J. Appl. Phys., 48:4991 (1977).
13. Yu. I. Bychkov, Yu. D. Korolev, and G.A. Mesyats, Pulse Discharges in Gases Under Conditions of Strong Ionization by Electrons, Sov. Phys. Usp., 21:944 (1978).

14. R.F. Fernsler, D. Conte, and I.M. Vitkovitsky, Repetitive Electron Beam Controlled Switching, *IEEE Trans. Plasma Science*, PS-8:176 (1980).
15. L.E. Kline, Performance Predictions for Electron-Beam Controlled On/Off Switches, *IEEE Trans. Plasma Sci.*, PS-10:224 (1982).
16. G. Schaefer, K.H. Schoenbach, P. Tran, J. Wang, and A.H. Guenther, Computer Calculations of the Time Dependent Behavior of Diffuse Discharge Switches, *Proc. 4th IEEE Pulsed Power Conf.* (1983).
17. K.H. Schoenbach, M. Kristiansen, E.E. Kunhardt, L.L. Hatfield, and A.H. Guenther, Exploratory Concepts of Opening Switches, *Proc. ARO Workshop on Repetitive Opening Switches*, Tamarron, CO, 65 (1981). DTIC No. AD-A110770.
18. L.G. Christophorou, S.R. Hunter, J.G. Carter, and R.A. Mathis, Gases for Possible Use in Diffuse-Discharge Switches, *Appl. Phys. Lett.*, 41:147 (1982).
19. W.L. Morgan, 1983, unpublished.
20. G. Schaefer, G.F. Reinking, and K.H. Schoenbach, Monte Carlo Calculations on the Influence of Attachment on the Thermalization of Secondary Electrons in an Electron-Beam-Sustained Discharge, *J. Appl. Phys.*, 61(2):790 (1987).
21. M.J. Berger, and S.M. Seltzer, Tables of Energy-Losses and Ranges of Electrons and Positions, *NAS-NRC Publication 1133*, 205 (1964).
22. ICRU Report 37, "Stopping Powers for Electrons and Positrons, International Commission on Radiation Units and Measurements," ICRU Publications, 7910 Woodmont Ave., Suite 1016, Bethesda, MD, 20814 (1984).
23. D.C. Lorents, and R.E. Olson, Excimer Formation and Decay Processes in Rare Gases, *SRI Semiannual Tech. Rep. No. 1*, Stanford Research Institute (1972).
24. S.R. Hunter, J.G. Carter, J.L. Christophorou, and V.K. Lakdawala, Transport Properties and Dielectric Strengths of Gas Mixtures for Use in Diffuse Discharge Opening Switches, *Gaseous Dielectrics IV*, Pergamon Press, New York (1984).
25. A.D. Barkalov, and G.G. Gladush, Domain Instability of a Non-Self-Sustaining Discharge in Electronegative Gases, I. Numerical Calcultions, *High Temp.*, 20:16 (1982); II. Theoretical Analysis, *High Temp.*, 20:172 (1982).
26. P. Bletzinger, Electron Beam Switching Experiments in the High Current Gain Regime, *Proc. 3rd IEEE Intern. Pulsed Power Conf.* (1981).
27. P. Bletzinger, E-beam Experiments--Interpretations and Extrapolations, *Proc. ARO Workshop on Diffuse Discharge Opening Switches*, Tamarron, CO, 112 (1982), DTIC No. AD-A115883.

28. M.R. Hallada, P. Bletzinger, and W. Bailey, Application of Electron-Beam Ionized Discharges to Switches--A Comparison of Experiment with Theory, IEEE Trans. Plasma Sci., PS-10:218 (1982).
29. P. Bletzinger, Scaling of Electron Beam Switches, Proc. 4th IEEE Pulsed Power Conf. (1983).
30. Yu. D. Korolev, G.A. Mesyats, and A.P. Khuzeev, Electrode Phenomena Preceeding the Transition of Non-Self-Maintaining Diffuse Discharge to a Spark, Sov. Phys. Dokl., 25:573 (1980).
31. D.H. Douglas-Hamilton, Some Instabilities in Non-Self-Sustained Gas Discharges, Proc. US-FRG Joint Seminar on Externally Controlled Diffuse Discharges (1983).
32. R.A. Haas, Plasma Stability of Electric Discharges in Molecular Gases, Phys. Rev. A, 8:1017 (1973).
33. R.A. Haas, Stability of Excimer Laser Discharges, Applied Atomic Collision Physics, Volume 3: Gas Lasers, Academic Press, New York (1982).
34. W.L. Nighan, Stability of High-Power Molecular Laser Discharges, Principles of Laser Plasmas, John Wiley & Sons, New York (1976).
35. E.E. Kunhardt, Basic Topics of Current Interest to Switching for Pulsed Power Applications, Gaseous Dielectrics IV, Pergamon Press, New York (1983).
36. D.H. Douglas-Hamilton and S.A. Mani, Attachment Instability in an Externally Ionized Discharge, J. Appl. Phys., 45:4406 (1974).
37. L.C. Pitchford, Power Loading Effects in Gas Discharge Models, Proc. US-FRG Joint Seminar on Externally Controlled Diffuse Discharges, 148 (1983).
38. A.P. Napartovich, V.G. Naumov, and V.M. Shashkov, Heating of Gas in a Combined Discharge in a Flow of Nitrogen, Sov. Phys. Dokl., 22:35 (1977).
39. J.A. Shirley, and R.J. Hall, Vibrational Excitation in H_2 and D_2 Electric Discharges, J. Chem. Phys., 67:2419 (1977).
40. J.N. Bardsley, and J.M. Wadehra, 1982, private communication.
41. J.M. Wadehra, Effect of Rot-Vibrational Excitation on Dissociative Attachment of H_2, 35th Gaseous Elec. Conf., (1982).
42. H. Seguin, J. Tulip, Photoinitiated and Photosustained Laser, Appl. Phys. Lett., 21:414 (1972).
43. D.R. Suhre, UV Sustained CO Laser Discharge, I. Photoionization Studies, J. Appl. Phys., 52:3855 (1981), and D.R. Suhre, and S.A. Wutzke, UV Sustained CO Laser Discharge, II. Discharge Studies, J. Appl. Phys., 52:3858 (1981).
44. W.M. Moeny, and M. von Dadelszen, UV-Sustained Glow Discharge Opening Switch Experiment, Proc. 5th IEEE Pulsed Power Conf., 630 (1985).

45. C.M. Young, M. von Dadelszen, J.M. Elizondo, J.W. Benze, M.W. Moeny, and J. G. Small, Investigation of a UV-Sustained Radial Glow Discharge Opening Switch, Proc. 5th IEEE Pulsed Power Conf., Arlington, VA (1985).
46. G.Z. Hutcheson, J.R. Cooper, E. Strickland, G. Schaefer, and K.H. Schoenbach, A UV-Sustained Discharge for Opening Switches, to be published (1987).
47. G. Schaefer, P.F. Williams, K.H. Schoenbach, and J.T. Mosley, Photodetachment as a Control Mechanism for Diffuse Discharge Switches, IEEE Trans. Plasma Sci., PS-11:263 (1983).
48. M.J. Rossi, H. Helm, and D.C. Lorents, Photoenhanced Electron Attachment in Vinylchloride and Trifluorethylene at 193 nm, Appl. Phys. Lett., 47:576 (1985).
49. F.L. Eisele, "Photon Induced Electron Attachment - Final Report," AFWAL-TR-85-2015 (1984).
50. S.J. Smith, Photodetachment Cross Section for the Negative Ion of Atomic Oxygen, Proc. 4th Int. Conf. Ionization Phenomena in Gases, IC:219 (1960).
51. J.L. Moruzzi, and A.V. Phelps, Survey on Negative-Ion-Molecule Reactions in O_2, CO_2, H_2O, CO, and Mixtures of These Gases at High Pressures, J. Chem. Phys., 45:4617 (1966).
52. L.M. Branscomb, S.J. Smith, and G. Tisone, Oxygen Metastable Atom Production Through Photodetachment, J. Chem. Phys., 43:2906 (1965).
53. D.S. Burch, S.J. Smith, and L.M. Branscomb, Photodissociation Spectroscopy of O_2, J. Chem. Phys., 112:171 (1958).
54. S.K. Srivastava, and O.J. Orient, A Double E-beam Technique for Collision Studies from Excited States: Application to Vibrationally Excited CO_2, Phys. Rev. A, 27:1209 (1983).
55. M. McFarland, D.L. Albritton, F.C. Fehsenfeld, E.E. Ferguson, and A.L. Schmeltekopf, Flow-Drift Technique for Ion Mobility and Ion-Molecule Reaction Rate Constant Measurements, III. Negative Ion Reactions of O^- with CO, NO, H_2, and D_2, J. Chem. Phys., 59:6629 (1973).
56. W.T. Whitney, and H. Hara, Improved Operation of N_2O TE Lasers, IEEE J. Quantum Electron., QE-19:1616 (1983).
57. L.G. Christophorou, "Electron-Molecule Interactions and Their Applications," Academic Press, Inc., Orlando (1984).
58. W.E. Wentworth, R. George, and H. Keith, Dissociative Thermal Electron Attachment of Some Alphatic Chloro, Bromo, Iodo Compounds, J. Chem. Phys., 51:1791 (1969).
59. K.J. Nygaard, H.L. Brooks, and S.R. Hunter, Negative Ion Production Rates in Rare Gas Halide Lasers, IEEE J. Quantum Electron., QE-15, 1216 (1979).
60. J.W. Gallagher, E.C. Beaty, J. Dutton, and L.C. Pitchford, A Compilation of Electron Swarm Data in Electro-Negative Gases, JILA Info. Center Report No. 22 (1982).

61. M. Allan, and S.F. Wong, Dissociative Attachment from Vibrationally and Rotationally Excited HCl and HF, J. Chem. Phys., 74:1687.
62. C.L. Chen, and P.J. Chantry, Photo-Enhanced Dissociative Electron Attachment in SF_6 and Its Isotopic Selectivity, J. Chem. Phys., 71:3897 (1979).
63. R. Barbe, J.P. Astruc, A. Lagreze, and J.P. Schermann, Spectral and Temperature Dependence of Laser Induced Dissociative Attachment in SF_6, Laser Chem., 1:17 (1982).
64. J.J. Tiee, and C. Wittig, Optically Pumped Molecular Lasers in the 11-17 Micron Region, J. Appl. Phys., 49:61 (1978).
65. E.R. Sirkin, and G.C. Pimentel, HF Rotational Laser Emission Through Photoelimination from Vinyl Floride and 1,1-Difluorothene, J. Chem. Phys., 75:604 (1981).
66. M.J. Berry, Chlorethylene Photochemical Lasers: Vibrational Energy Content of HCl Molecular Elimination Products, J. Chem. Phys., 61:3114 (1974).
67. C.R. Quick, and C. Wittig, IR Multiple Photon Dissociation of Fluorinated Ethanes and Ethylenes: HF Vibrational Energy Distributions, J. Chem. Phys., 72:1964 (1980).
68. P.L. Houston, Electronic to Vibrational Energy Transfer from Excited Halogen Atoms, in Photoselective Chemistry--Part 2, John Wiley and Sons, New York (1981).
69. A.B. Peterson, G. Wittig, and S.R. Leone, Electronic to Vibrational Pumped CO_2 Laser Operating at 4.3, 10.6, and 14.1 μm, J. Appl. Phys., 47:1057 (1976).
70. I.M. Beterov, and N.V. Fatayev, Optogalvanic Demonstration of State-to-State Dissociative Electron Capture Rate in I_2, Opt. Comm., 40:425 (1982).
71. J.E. Eninger, Broad Area Electron Beam Technology for Pulsed High Power Gas Lasers, Proc. 3rd IEEE Pulsed Power Conf., 499 (1981).
72. T.J. Orzechowski, and G. Bekefi, Current Flow in a High-Voltage Diode Subjected to a Crossed Magnetic Field, Phys. Fluids, 19:43 (1976).
73. C.A. Spindt, I. Brodie, L. Humphrey, and E.R. Westerberg, Physical Properties of Thin-Film Field Emission Cathodes with Molybdenum Cones, J. Appl. Phys., 47:5248 (1976).
74. W.M. Clark, "Ion Plasma Electron Gun Research," HRL Final Report, ONR Contract N00014-77-C-0484 (1977).
75. G. Wakalopulos, and L. Gresko, "Pulsed WIP Electron Gun, Final-Report--Design Phase," HAC Report No. FR-79-73-735, UCRL (1979).
76. H.E. Gallagher, and R.J. Harvey, Development of a 1-GW Electron-Beam Controlled Switch, Proc. 16th Power Modulator Symp., 158 (1984).

77. C.H. Harjes, K.H. Schoenbach, G. Schaefer, M. Kristiansen, H. Krompholz, and D. Skaggs, Electron-Beam Tetrode for Multiple, Submicrosecond Pulse Operation, Rev. Sci. Instrum., 55:1684 (1984).
78. R. Petr and M. Gundersen, Field Emission Cathode for High Current Beams, Lasers and Particle Beams, 1:207 (1983).
79. J.L. Cronin, Modern Dispenser Cathodes, Proc. IEE, 128:19 (1982).
80. J.D. Daugherty, Electron Beam Ionized Lasers, Principles of Laser Plasmas, G. Bekefi, ed., John Wiley & Sons, Inc., New York (1976).
81. D.H. Douglas-Hamilton, Diffuse Discharge Production--Group Report, Proc. ARO Workshop on Diffuse Opening Switches, 169 (1982) DTIC No. AD-A115883.
82. R.J. Commisso, R.F. Fernsler, V.E. Scherrer, and I.M. Vitkovitsky, Higher Power Electron Beam Controlled Switches, Rev. Sci. Instrum., 55:1834 (1984).
83. Yu. I. Bychkov, Yu. D. Korolev, G.A. Mesyats, A.P. Khuzeev, and I.A. Shemyakin, Volume Discharge Excited by an Electron-Beam in an Ar-SF_6 Mixture, I. Beam Controlled Discharge, Izvestiya Vysshikh Uchebnykh Zavendenii, 7:898 (1979); II. Cascade Ionization Discharge, Izvestiya Vysshikh Uchebnykh Zavendenii, 7:77 (1979).
84. A.M. Efremov, and B.M. Koval'chuk, A Non-Self-Consistent Electron-Beam Controlled Discharge in Methane, Izvestiya Vysshikh Uchebnykh Zavedinii, 4:65 (1982).
85. L.C. Lee, and F.Li, Shortening of Electron Conduction Pulses by Electron Attachers, O_2, N_2O, and CF_4, J. Appl. Phys., 56:3169 (1984).
86. K.H. Schoenbach, G. Schaefer, M. Kristiansen, H. Krompholz, D. Skaggs, and E. Strickland, An E-Beam Controlled Diffuse Discharge Switch, Proc. 5th IEEE Pulsed Power Conf., 640 (1985).
87. L.C. Lee, C.C. Chiang, K.Y. Tang, D.L. Huestis, and D.C. Lorents, Gaseous Electronic Kinetics for E-Beam Excitation of Cl, NO and N_2O in N_2, Second Annual Rep. on AFOSR Coord. Res. Program in Pulsed Power Physics, Dept. Elec. Engineering, Texas Tech Univ., Lubbock, TX, 189 (1981).
88. J.F. Lowery, L.E. Kline, and J.V.R. Heberlein, Electron-beam Controlled On/Off Discharge Characterization Experiments, Proc. 4th IEEE Pulsed Power Conf., 94 (1983).
89. R.J. Commisso, R.F. Fernsler, V.E. Scherrer, and I.M. Vitkovitsky, Electron-Beam Controlled Discharges, IEEE Trans. Plasma Sci., PS-10:241 (1982).
90. R.J. Commisso, R.F. Fernsler, V.E. Scherrer, and I.M. Vitkovitsky, Application of Electron-Beam Controlled Diffuse Discharges to Fast Switches, Proc. 4th IEEE Pulsed Power Conf., 87 (1983).

91. V.E. Scherrer, R.J. Commisso, R.F. Fernsler, L. Miles, and I.M. Vitkovitsky, The Control of Breakdown and Recovery in Gases by Pulsed Electron Beams, Gaseous Dielectrics III, Pergamon Press, New York (1982).
92. V.E. Scherrer, R.J. Commisso, R.F. Fernsler, and I.M. Vitkovitsky, Study of Gas Mixtures for E-beam Controlled Switches, in Gaseous Dielectrics IV, Pergamon Press, New York (1984).
93. K.H. Schoenbach, G. Schaefer, M. Kristiansen, H. Krompholz, H. Harjes, and D. Skaggs, Investigations of E-Beam Controlled Diffuse Discharges, Gaseous Dielectrivs IV, Pergamon Press, New York (1984).
94. K.H. Schoenbach, G. Schaefer, M. Kristiansen, H. Krompholz, H. Harjes, and D. Skaggs, An E-Beam Controlled Diffuse Discharge Switches, Proc. 16th Power Modulator Symp., 152 (1984).
95. K.H. Schoenbach, G. Schaefer, M. Kristiansen, H. Krompholz, H. Harjes, and D. Skaggs, An Electron-Beam Controlled Diffuse Dischage Switch, J. Appl. Phys., 57:1618 (1985).
96. G. Schaefer, K.H. Schoenbach, M. Kristiansen, B.E. Strikland, R.A. Korzekwa, and G.Z. Hutcheson, Influence of the Circuit Impedance on an Electron Beam Controlled Diffuse Discharge with a Negative Differential Conductivity. Appl. Phys. Lett., 48:1776 (1986).
97. R.J. Commisso, R.F. Fernsler, V.E. Scherrer, and I.M. Vitkovitsky, Inductively Generated, High Voltage Pulse Using an Electron Beam Controlled Opening Switch, Appl. Phys. Lett., 47:1056 (1985).
98. A.H. Guenther and J.R. Bettis, Recent Advances in Optically Controlled Discharges, Proc. 5th IEEE Pulsed Power Conf., 47 (1985).

LOW-PRESSURE PLASMA OPENING SWITCHES

Robert W. Schumacher and Robin J. Harvey

Plasma Physics Department
Hughes Research Laboratories
Malibu, CA 90265

INTRODUCTION

Plasma switch devices such as thyratrons, spark gaps, and ignitrons have enjoyed great success in closing-switch applications where high currents must be switched from high voltage on short time scales with low forward-voltage drop. However, plasma devices are not usually considered as repetitive fast opening switches. Although they can be opened by commutation (or by counterpulse techniques), commutation requires significant power, and recovery times are generally long: $\gtrsim 10$ μs. Ideally, a plasma switch is desired which can perform both closing and opening functions at high speed and high repetition rates under low-power control, like a hard-vacuum tube, but without the large forward-voltage drop and low thermionic-cathode current restrictions which are associated with hard tubes.

Low-pressure, diffuse-discharge plasma switches retain many of the characteristics required for this "ideal" operation. The low-pressure switch employs a high-voltage gap which operates on the left-hand side of the Paschen discharge curve to hold off high voltage up to 200 kV in the open state against both vacuum and Paschen breakdown mechanisms. The switch is closed by injecting plasma into the gap, or by applying a magnetic field which traps electrons, thus increasing the effective gap spacing and thereby triggering a Paschen or glow discharge. Current interruption is initiated by terminating the injection of plasma or by turning off the applied magnetic field. Since the ambient gas pressure is too low to sustain ionization in the gap, the plasma decays and allows the switch to open on an ion-diffusion time scale.

Although this type of pulsed-power switch was first tested in 1946,[1] low-pressure plasma opening-switch technology is only now maturing to the point where reliable, long-life, vacuum-sealed tubes are being developed. Most of the work in the repetitive plasma-opening-switch area has been performed at Hughes Research Laboratories (HRL), a division of Hughes Aircraft Company. <u>One exception</u> is the Tacitron switch; a hot-cathode thyratron-like device which was originally conceived and tested at RCA,[2] and later developed to commercialization at peak power levels up to 3.6 MW in the Soviet Union.[3]

Hughes has developed two low-pressure, plasma opening switches: the magnetically switched Crossed-Field Tube[4-6] (XFT) and the electrostatically controlled CROSSATRON™ Modulator Switch[7,8] (CMS). Both switches are secondary-electron-emitter, cold-cathode devices which employ a controlled diffuse discharge to both close and open pulsed-power circuits at high speed and at high repetition frequency. Vacuum-sealed-prototype switch tubes have been developed at power levels up to 1 GW using the XFT approach, and up to 20 MW using the CMS design. In comparison to conventional dc-current opening-switch devices such as hard-vacuum tetrodes, these two low-pressure plasma switches offer several significant advantages:

· No cathode heater
· Instant starting
· Long life
· Low forward-voltage drop
· High current conduction
· Electro-mechanically rugged operation.

These important features make the low-pressure plasma switch an attractive device. The following sections summarize the operating principles, the current state of the art, and the ultimate capabilities of low-pressure plasma opening switches.

BASIC PHYSICS OF LOW-PRESSURE, GAS-DISCHARGE OPENING SWITCHES

Low-pressure plasma switches are simply diffuse glow-discharge[9] tubes which operate on the low-pressure side of the Paschen breakdown curve. On-off switching is controlled by an externally-applied magnetic field (XFT) or by modulating the injection of remotely generated plasma into the low-pressure switching gap (CMS). As implied by the term "glow discharge," these switches are cold-cathode devices which employ secondary-electron emission techniques to sustain the gas discharge which conducts the switch current. In this section we review the physics of the glow discharge and the limitations it imposes on hold-off voltage, current density and forward-voltage drop.

Paschen Breakdown: The Basis of a Switch

Paschen breakdown is the term usually used to describe a diffuse, glow discharge through a gas between two high-voltage electrodes. The breakdown voltage (Figure 1) is a function of the pressure-times-electrode-spacing product, or Pd. For most gases the breakdown curve has a minimum at about 1 Torr-cm and rises sharply on the left-hand side where Pd is low.[10] Under the Paschen curve in this region, the ionization mean free path is long compared to the gap spacing so that stray electrons in the gas are collected at the electrodes before they can initiate an avalanche. Under the right-hand side of the curve in Figure 1, where the Pd product is large, the mean free path is very short so electrons do not gain sufficient energy between collisions to cause ionization and subsequent breakdown.

Since the breakdown voltage is a strong function of Pd, a gas-filled high-voltage gap operating on either side of the curve forms the basis of an effective electrical switch. For example, if the gap is operated at low Pd to the left of the curve in Figure 1, high voltage can be held off across the electrodes without conduction. This is the OPEN state of the switch. The switch can be closed and thereby made to conduct current between the electrodes by either injecting plasma into the gap from the outside, or by suddenly increasing the Pd product such that the position marked CLOSED is reached, Paschen breakdown is initiated, and the self-ionized gas conducts current. In practice, Pd is increased by

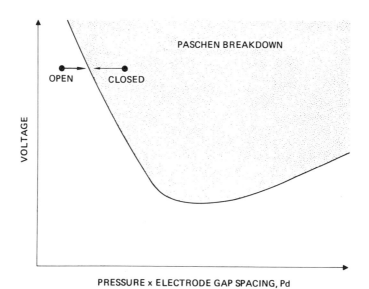

Fig. 1. Paschen breakdown curve.

applying a magnetic field which traps electrons in the gap and significantly increases the effective gap size. To re-open the switch, plasma injection is terminated or the Pd value is reduced to the OPEN position. At this point the gas pressure is too low to sustain ionization, so the plasma in the gap begins to decay. As the plasma density decreases, the gap impedance rises, the conduction current falls, and the gap voltage rises toward the open-circuit value. The exact risetime of the gap voltage depends not only upon the plasma decay time but also upon the external circuit impedance and the stray capacitance across the gap.

In theory the gap could be operated as an opening switch on either side of the Paschen curve. But in practice it is generally more effective to operate on the low-pressure side. Here, the electron-neutral collision rate is lower and the electron temperature is higher, which leads to a more rapid current interruption and a faster gap-voltage recovery time. Because the gap is operated at such low (> 50 mTorr) pressure, the ionization mean free path is large compared to the gap spacing. Consequently, the physics of the gap plasma is dominated by collisionless, collective phenomena rather than by collisional effects. For example, the current interruption time is very nearly equal to the plasma ambipolar-diffusion[11] time in the gap, which is approximately

$$\tau_d = \frac{d}{\sqrt{\frac{kT_e}{M_i}}}, \qquad (1)$$

where T_e is the electron temperature, M_i is the ion mass, and k is the Boltzmann constant. For typical values (~ 1 cm) of inter-electrode gaps, and electron temperatures ~ 5 eV, this equation predicts opening times of 0.5 to 5 μs, which are consistent with experimental data.

Vacuum-Breakdown Limitations

Unlike the inverse practice at high pressure, one cannot simply decrease the value of Pd at low pressure to obtain arbitrarily higher open-circuit voltage. Eventually, the maximum hold-off voltage becomes limited by field emission (at the cathode electrode) which triggers vacuum breakdown.[12] As is the case for Paschen breakdown, the actual vacuum-breakdown voltage level depends not only upon the gap spacing but also upon the electrode geometry, electrode material, electrode surface condition, and the external circuit. The vacuum-breakdown curve shown in Figure 2, which plots breakdown voltage against gap spacing, represents about the best performance one can expect with carefully polished and discharge-conditioned electrodes which are made from ultra-clean,

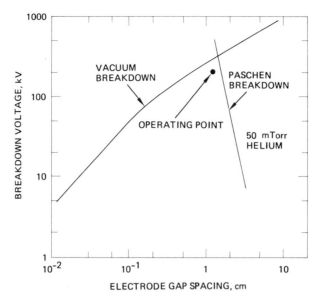

Fig. 2. The operating point of single-gap, low-pressure plasma switches is determined by the cross-over of vacuum and Paschen breakdown curves.

inclusion-free materials such as titanium, stainless steel, or molybdenum.

The highest open-circuit voltage attainable across a single-gap low-pressure plasma switch is approximately 200 kV. Above this voltage, the Paschen breakdown curve for helium gas at 50 mTorr intersects the vacuum breakdown curve, as shown in Figure 2. Since helium has the highest ionization potential and the smallest ionization cross-section of any gas, it provides the highest hold-off voltage against Paschen breakdown. Consequently, the operating point (Figure 2) which can be chosen for a low-pressure plasma switch is located just under the cross-over of the vacuum and Paschen breakdown curves and provides a maximum open-circuit voltage of ~ 200 kV across a 1.3-cm gap. Although a marginal increase in hold-off voltage may be obtained by operating at lower gas pressures, the switch becomes more difficult to close[13] and also closes at a slower rate (lower dI/dt) at P < 50 mTorr.

Glow-Discharge Maintenance

Once the glow discharge is initiated in a low-pressure switch, the discharge is maintained by secondary-electron emission from the cold cathode. Figure 3 plots the steady-state, glow-discharge potential distribution between the cathode electrode on the left and the anode electrode on the right. The potential of the plasma

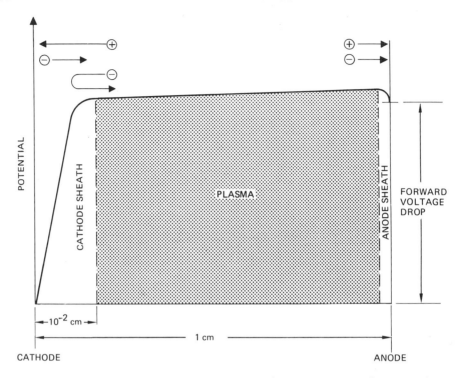

Fig. 3. Potential distribution in a low-pressure glow-discharge gap.

relative to the cathode is generally 200 to 1000 V, depending upon the gas species and electrode materials used, as well as the current density at the cathode. Ions are collected from the plasma in the gap across non-neutral sheath regions at both the cathode and anode. Electrons, however, are collected at the anode electrode only. Here, the plasma maintains a small anode-sheath voltage drop on the order of several kT_e/e to adjust the ambipolar flux of electrons and ions such that the plasma remains electrically neutral. The behavior is just the opposite of that expected in the high-pressure regime where the anode-sheath voltage increases by an amount which is on the order of the ionization potential of the gas. Most of the potential drop across the switch occurs at the cathode sheath where ions are accelerated to kinetic energy levels sufficient to stimulate the emission of secondary electrons from the cathode surface.

The total cathode current is therefore the sum of the ion current collected from the plasma and the emitted secondary-electron current, or

LOW-PRESSURE PLASMA OPENING SWITCHES

$$I_c = I_i + \gamma_e I_i . \quad (2)$$

In this equation γ_e is the secondary-electron yield (number of electrons emitted per incident ion) which ranges from a few percent to values in excess of unity, depending upon cathode material, cathode-surface condition, ion mass, ionization state of the gas, and the cathode-sheath voltage drop. Since the ion current col-

$$J_i = \frac{4\epsilon_o}{9}\sqrt{\frac{2e}{M_i}}\frac{V_c^{3/2}}{(\Delta x)^2}, \quad (3)$$

lected by the cathode is space-charge limited, the cathode-sheath thickness, Δx, is determined by the Child-Langmuir law, where J_i is the ion current density drawn from the plasma and V_c is the cathode-sheath voltage drop. For a helium ion flux of 10 A/cm^2 falling through a 500-V sheath drop, the sheath thickness is on the order of 10^{-2} cm, or about 1 percent of the gap spacing. Equation (3) accurately describes the cathode-fall characteristics since all collision mean-free paths are $\gg \Delta x$ and since the plasma is weakly magnetized such that ion and high-energy-electron gyroradii are $\gtrsim \Delta x$.

Following their emission from the cathode, the secondary electrons are accelerated back through the cathode sheath and enter the plasma at an energy corresponding to the 200- to 1000-V cathode-sheath drop. If a magnetic field is present to trap these electrons between the electrodes, then the high-energy electrons will undergo ionizing collisions with the background neutral-gas atoms in the gap before they are collected by the anode. In the steady state, the rate of ionization balances the ion loss rate to the electrodes such that the glow-discharge plasma is maintained at a constant density in time.

The equilibrium plasma density can be computed by evaluating an equation which balances the ion generation and loss rates in a weakly ionized plasma (plasma density $\ll$ neutral gas density):

$$\frac{NI_{se}}{e} = \frac{n_i A d}{\tau_d}. \quad (4)$$

In this equation, I_{se} is the secondary electron current injected into the plasma, N is the average number of ionizing collisions made by a secondary electron before it is collected by the anode or before its energy drops below the ionization potential of the gas, n_i is the ion density, A is the electrode area, d is the gap spacing, and τ_d is the ion ambipolar diffusion time given by Eq. (1). Using Eq. (2), I_{se} can be expressed in terms of γ_e and the total cathode current as

$$I_{se} = \frac{\gamma_e}{1+\gamma_e} I_c . \tag{5}$$

Equation (4) can then be solved for the ion density using Eq's. (1) and (5) to obtain

$$n_i = \frac{N J_c}{e} \frac{\gamma_e}{1+\gamma_e} \left(\frac{kT_e}{M_i} \right)^{-1/2} , \tag{6}$$

where J_c is the total cathode current density. This equation shows that the plasma density in the gap scales linearly with the cathode current density and with the average number of ionizing collisions made by the secondary electrons. Thus, the plasma density increases as the gas pressure rises or as the electron trapping efficiency improves (both of which increase N). For a switch gap conducting 10 A/cm² in 50 mTorr of helium with T_e = 5 eV, the plasma density is approximately 10^{14} cm⁻³. Since the gas density at this pressure is about 10^{15} cm⁻³, the gas is ~ 10 percent ionized. Other mechanisms, such as beam-plasma instabilities, metastable excitation, and charge exchange may influence the accuracy of Eq. (6), but typically, low-pressure switches employ such weakly ionized plasmas.

Forward-Voltage Drop

In a glow discharge the forward-voltage drop is self-regulated in the 200- to 1000-V range because of the low values of γ_e, which are generally encountered when helium or hydrogen discharges are run with electrode materials such as stainless steel, titanium, or molybdenum. With a typical γ_e of 30 percent,[14] for example, the cathode will require three incident ions before a secondary electron is emitted. To maintain a steady discharge, this electron must make at least three ionizing collisions to replace the three ions which were lost to the cathode. Since the ionization energy of helium, for example, is 25.6 eV, the discharge voltage must be at least 77 V in order to provide enough energy for the three ionization events. In an actual glow discharge under these conditions, the voltage would be much higher to make up for ion losses at locations other than the cathode (such as at the anode, grids, or end plates) and for other secondary-electron energy losses such as excitation of metastable atoms, optical line radiation from non-ionizing collisions, and beam-plasma instabilities.[15]

The secondary yield, γ_e, is a sensitive function of not only the ion species but also the chemical composition and gas loading of the cathode surface. An important feature is that γ_e also increases with ion energy. For this reason, the discharge voltage will adjust to reach a stable equilibrium which may vary considerably from device to device. The parameter N is likewise influenced by the device design.

As the cathode current density increases above 0.1 A/cm^2, the magnitude of the forward-voltage drop generally increases in order to supply the additional current. This regime of the diffuse discharge is customarily called the "abnormal glow[9]." In Fig. 4 the forward-voltage drop across a helium discharge with stainless-steel electrodes is plotted as a function of cathode current density. The voltage increases from 200 V at 0.1 A/cm^2 to over 1000 V above 10 A/cm^2.

Although the higher voltage drop at high current density may be deemed undesirable from the standpoint of power dissipation and cathode heating, the positive coefficient of forward-voltage increment with current shown in Fig. 4 insures that multiple low-pressure switches connected in parallel will share current equally without requiring additional circuit elements. If two switches are connected in parallel, for example, and one switch begins to carry more current than the other, its voltage drop will increase and force the additional current to be shared by the parallel tube having a lower drop. This feature has important consequences since it suggests that a large parallel array of identically manufactured

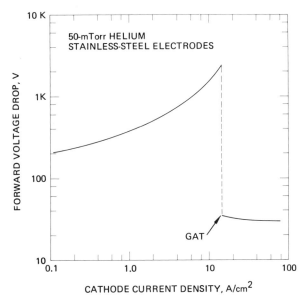

Fig. 4. Forward-voltage drop across a low-pressure glow discharge as a function of cathode current density. GAT represents the glow to arc transition.

low-pressure switches can be assembled to switch very high currents (≥ 100 kA).

Glow-To-Arc Transition

The glow discharge employed by low-pressure plasma switches is characterized by a diffuse distribution of the current over the switch-electrode area and a self-regulated discharge voltage of 200 to 1000 V. As the cathode current density rises to values exceeding (typically) 10 A/cm^2, however, the glow discharge may make a transition to an arc discharge which is characterized by a low discharge voltage (as shown in Fig. 4) and a concentration of all the switch current to a cathode arc spot. When this glow-to-arc transition[16,17] (GAT) occurs, the cathode-emission mechanism changes from secondary-electron emission to thermionic emission from a vapor arc where all the vapor necessary to sustain the arc is provided by the eroding cathode surface. The arc is therefore self-sustaining as long as current flows through the switch. Although a typical arc of reasonably limited current and duration does not damage the switch, control of the discharge is lost once the GAT occurs since the arc will maintain conduction even in the absence of magnetic-field trapping or plasma injection.

Reliable operation of low-pressure repetitive plasma-opening switches therefore requires that the cathode current density not exceed 10 A/cm^2, particularly during current interruption. In practice, the current density on well-conditioned electrodes may exceed this limit during switch conduction without GAT. But when current interruption is attempted the probability of arc formation is substantially increased.

Glow-to-arc transition can be triggered by several different mechanisms. One is field-emission-induced breakdown from cathode-surface micro-protrusions in the high electric field of the cathode sheath. From Eq. (3) it can be shown that during conduction the sheath electric field in the absence of protrusions is $\leq 10^7$ V/m and scales as the fourth root of the voltage across the switch (i.e., $E \sim J^{1/2} V^{1/4}$). During interruption, when the switch voltage is rising, the sheath field attains its maximum value. If the field at the tip of a micro-protrusion is already near threshold during conduction, then field emission and subsequent arc formation would occur when the switch is opened.

Another mechanism responsible for GAT is the concentration of plasma that may occur at interruption. This process is of particular concern in crossed-field tubes where interruption is initiated by turning-off the magnetic field. Since the metallic vacuum wall traps magnetic flux inside the tube, the internal magnetic field must diffuse through the vacuum wall before the field is reduced below the interruption level. The diffusion process is not perfectly uniform so there will be isolated regions on the cathode wall during interruption where the field strength is still sufficient to trap electrons and maintain the discharge. The plasma and the discharge current will tend to concentrate in these regions, leading to a local spot of increased current density which may trigger an arc. This mechanism is of less concern in the CROSSATRON Modulator Switch (described later) where current interruption is performed electrostatically in a more uniform manner.

Finally, there is also evidence that GAT is triggered by insulating inclusions and micro-particles (such as dust) on the cathode surface.[17] These structures are subject to significant electrical charging in the cathode-sheath electric field. As the field increases, the electrical force causes the particles to explode, thus producing the vapor burst required for an arc spot to be established on the cathode surface. Reliable operation at high current densities therefore requires the use of inclusion-free cathode materials and the employment of clean-room assembly techniques to avoid the introduction of dust or other micro-particles into the switch-tube envelope.

CROSSED-FIELD TUBES

Geometry

Crossed-Field Tubes (XFT) are low-pressure plasma-switch devices which employ a magnetic field to control conduction. As shown in Fig. 5, they are generally coaxial devices with a pulsed electromagnet wound on the outer cylinder which is usually the cathode electrode. The center electrode is the anode which is insulated from the cathode by a high-voltage feedthrough bushing. The outer electrode is preferred for the cathode because (1) it provides more cathode surface area to minimize the cathode current density and (2) because this polarity, combined with the curved magnetic field lines resulting from coils positioned outside the cathode, allows crossed-field breakdown (see next section) at a lower magnetic field.

An electron located in the coaxial gap between the two cylinders in Fig. 5 moves under the influence of the radially directed electric field and the axially directed magnetic field (pointing out of the page in Fig. 5). In this crossed-field geometry the electron executes a well-defined trajectory which drifts in the azimuthal, or $\bar{E} \times \bar{B}$, direction and in the axial direction along the magnetic field. Since the cylinders must be of finite length, a short solenoid or cathode potential end plates are employed in order to limit the axial drift; either by directing the electrons to the cathode or by electrostatically reflecting electrons from the ends of the tube. This crossed-field confinement geometry

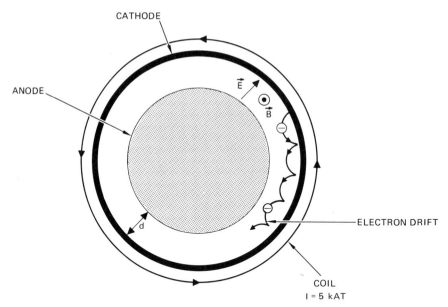

Fig. 5. Geometry of the crossed-field tube.

efficiently confines electrons until they make collisions with background gas atoms or molecules.

Ignition (ON-Switching)

A XFT is normally ignited in the following manner. First, the voltage is applied to the tube in the absence of any magnetic field; the magnetic field required for ignition is then applied. It can be shown that the field required to trap electrons varies with the square root of the applied voltage according to[18]

$$B = \frac{1}{d}\sqrt{\frac{2mV}{e}}, \quad (7)$$

where m and e are the electronic mass and charge, respectively, d is the inter-electrode spacing (<< cathode radius), and V is the applied voltage. At V = 60 kV and d = 10^{-2}m, the field required for ignition is 8.3 x 10^{-2} Tesla (830 Gauss). Eq. (7) is plotted in Fig. 6.

The theoretical value of magnetic field calculated above is the field which must be present in the interelectrode space for ignition to occur. For typical XFT coils and pulsers, about 5 kilo-Ampere-turns (5 kAT) are required to reach the internal field

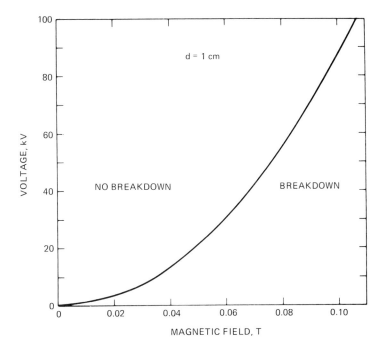

Fig. 6. Crossed-field breakdown behavior

necessary for ignition due to eddy currents in the vacuum envelope and the electrodes.

Once the breakdown conditions are satisfied, ignition will not occur unless some initial electrons are present to start an avalanche. Without any active means for providing these electrons, one must rely on cosmic rays or field emission from micro-protrusions on the cathode surface. The former are too infrequent and the latter depends on the applied voltage. Because applications typically require jitter times < 1 μs which are independent of applied voltage, a pre-ionization technique is usually employed. One technique is to use an auxiliary ionizer electrode[19] to provide field-emitted electrons at a constant rate from an independent power source. A tungsten pin biased a few kilovolts negative with respect to its anode with a current flow of a few microamperes is sufficient for this purpose. Another technique is to use a wire-ion-plasma (WIP) discharge[20] in a small appendage chamber to provide the seed electrons. The WIP device employs a glow discharge to a fine-wire anode.[21] A low-density glow discharge is established with only a few kilovolts applied to the (positive) wire as a result of electrostatic trapping of electrons in circular orbits about the wire.

Conduction

Once ignition has occurred and the anode voltage has fallen to a low value, the tube enters the conduction phase. As discussed above, the discharge voltage is in the range of 200 to 1000 V, depending on cathode material, end configuration, pressure, and magnetic field strength. Current densities > 10 A/cm^2 can be reliably conducted in properly processed and conditioned XFTs. The current density is limited by the glow-to-arc transition. A typical arc does no permanent damage to the tube, but once struck, removal of the magnetic field does not interrupt the current.

Interruption

When the magnetic field falls below the critical value required for sustaining the discharge, ionization ceases and the charge begins to be swept out of the tube. As with other electronic devices, a finite time (~5 to 10 μs) is required to complete this charge sweep-out. The dissipation which occurs during the interruption phase is the time integral of the current-voltage waveform across the XFT during interruption.

When the voltage rises across the XFT the circuit current falls. The power density (voltage times current density) at first rises and then falls. The peak power density can be high, and it is this peak value which can determine the maximum allowable current density that can be achieved without a high arc rate. Thus, the maximum operating current density depends upon the way in which the voltage rises when the XFT interrupts. This, in turn, depends in detail upon the circuit, its inductance, and any stray capacitance. (A similar situation prevails during ignition, but the anode fall time is an order of magnitude shorter than the interruption time, making this effect less important).

XFT Construction and Performance

Cross-field tubes have been constructed in a wide variety of sizes for both experimental purposes and for specific applications. These applications include dc-current-interrupter service for HVDC transmission lines,[22] repetitive-series-interrupter service for neutral-beam arc-protection circuits,[23] and service as an ac-current-limiting device in electric-utility systems.[24] Tube sizes have ranged from diameters of less than 5 cm to diameters of greater than 56 cm with active cathode areas ranging from 50 cm^2 to 8000 cm^2. The smaller tubes have nominal ratings for the opening mode starting at 50 kV and 280 A, while the larger tubes have operated with open-circuit voltages up to 120 kV and maximum interruptible currents up to 10 kA.

A schematic of a dual-anode XFT is shown in Fig. 7. It has a mean radius of 20.3 cm and a gap spacing of 1.5 cm. The anode is split at its center into two isolated sections which may be operated separately, each with its own magnetic field coils and high voltage feedthrough bushings. The tube is divided into two vacuum regions. The high vacuum region contains the electrodes. The rough (outer) vacuum region contains the magnetic field coils. The reason for this type of construction was to permit as thin a cathode wall as possible to minimize eddy currents and allow rapid magnetic-field penetration. The resulting wall thickness was not sufficient to withstand full atmospheric pressure, so the region outside the cathode was always "roughed out" at the same time as the high-vacuum section. This dual-anode device has nominal ratings of 100-kV open-circuit voltage and a maximum interruptible current of 8.2 kA. Typical current and voltage waveforms obtained using this switch in an inductive-energy-storage (IES) circuit are shown in Fig. 8. The 7.8-kA current is switched to essentially zero in about 5 microseconds. During this time, the load-resistor current and voltage rise approximately linearly to maxima of 5.1 kA and 102 kV. The energy transfer efficiency to the load resistor depends principally upon the size of the inductor, and in this case is 83 percent.

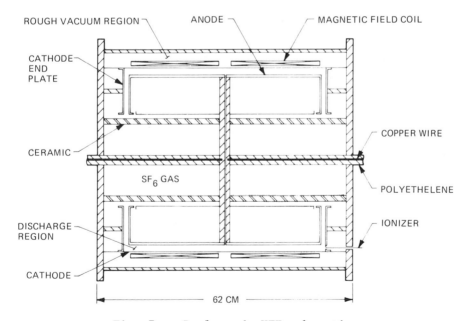

Fig. 7. Dual-anode XFT schematic.

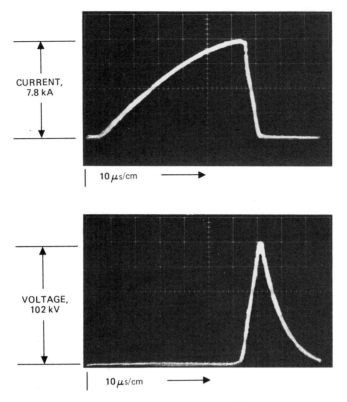

Fig. 8. Dual-anode XFT switching results in an IES circuit.

The largest XFT constructed to date is a single-ended "bipolar" device with a mean radius of 28 cm and a gap spacing of 1.9 cm. The electrodes are constructed of commercially pure (99+ percent) titanium spinnings. The tube is vacuum-sealed and is pressure controlled using a system consisting of a small ion pump and a helium leak valve. The alumina insulator (supporting the cathode and the anode), corona shields and interaction gap are designed to hold off 100 kV. The insulator length is 20 cm. As with the smaller tubes, a field emitter is installed at the base of the tube to provide an initial ionization source. Typically, the switch coil is wound about the center of the cathode and is about 20-cm wide. This large interrupter tube has a nominal power rating of 1 GW (100 kV, 10 kA).

Finally, a photograph of one of the smallest XFT devices is shown in Fig. 9. This tube was constructed for life-test purposes in an inductive-energy-storage circuit, and was operated at 50 kV open-circuit voltage, 150-A interruption current, and 80-Hz pulse repetition frequency for more than one million pulses without

LOW-PRESSURE PLASMA OPENING SWITCHES

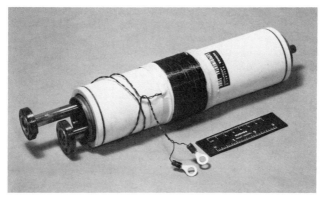

Fig. 9.   Photograph of a life-test-model (50 kV, 150 A), 50-cm^2 active-area XFT with internal molybdenum electrodes and a ceramic vacuum envelope.

failure. It is also a vacuum-sealed device which is pressure regulated using an ion pump and a helium leak valve. The 50-cm^2 active-cathode-area tube is constructed with internal, molybdenum electrodes and a ceramic vacuum envelope in the manner of modern hydrogen thyratrons. As shown in Fig. 9, the magnetic-field coil which controls the switch is wound about the mid-section of the tube.

CROSSATRON MODULATOR SWITCH

The CROSSATRON[25] Modulator Switch[26] (CMS) is a hybrid low-pressure plasma device which combines the best features of XFTs and gridded, vacuum-tube devices. Like the XFT, the CMS uses a low-pressure crossed-field discharge to generate a high-density plasma for conducting high currents with low forward drop. But, like a gridded vacuum tube, the CMS controls the conduction of electrons to the anode by controlling the potential of gridded electrodes which are located between the cathode and anode electrodes. As mentioned earlier, the high-voltage gap near the anode is always operated on the left-hand side of the Paschen breakdown curve, and switching is implemented by modulating the injection of plasma into the gap. The plasma is generated remotely by crossed-field breakdown near the cathode in a steady magnetic field produced by permanent magnets. To close the switch, the grids are driven positive with respect to the cathode and plasma is injected into the anode gap. To open the switch, the grids are returned to cathode potential or below, and plasma injection is terminated.

Although the CMS has not yet been developed to the current and voltage levels demonstrated with XFTs, high voltage current-interruption experiments have been performed at current density levels up to 7 A/cm^2 and total currents up to 500 A. Three switch devices have been constructed for these experiments. These include a 9.5-cm-diameter test-model CMS mounted on a vacuum station and operated with a controlled leak rate; a 15-cm-diameter "productized" switch which is vacuum sealed and operates with an internal titanium-hydride gas controller; and a 25-cm, 100-kV switch which has modulated average-power levels up to 1 MW in short bursts. Operation has been demonstrated as both a modulator switch with a resistive load and as an inductive-energy-storage (IES) circuit opening switch with an open circuit voltage up to 90 kV, conduction voltage of only 500 V, and opening times of 150 ns. The power required to initiate interruption in these experiments is negligible - a simple TTL level signal from a high impedance pulser is sufficient. At lower current, ~ 30 A, ultra-fast interruption times of ~ 50 ns have also been demonstrated with low jitter (~ 5 ns). In operation as a closing switch, this device has closed from 30 kV to conduct 1700 A with a 20-ns risetime at 16-kHz PRF. As a consequence of the fast recovery time ($\lesssim$ 1 μs at a current density $\leq$ 5 A/cm^2), CROSSATRON Modulator Switches are also capable of dual-pulse-modulator service with a short, variable dwell time between pulses. This feature has been used to produce two 2-μs-wide pulses at 15 kV and 45 A, with variable dwell times as short as 2 μs and with 200-ns rise and fall times.

In this section we review the basic principles and operating mechanisms of the CROSSATRON Modulator Switch. We begin with a brief discussion of the historical background of grid-controlled plasma-opening switches.

Historical Background of Grid-Controlled Plasma Opening Switches

Although there have been attempts to develop low-pressure gas-discharge switch devices which open through electrostatic control of grids immersed in a plasma, these efforts have failed (except in the USSR) to produce a commercial, high-power modulator switch. To our knowledge, the first successful electrostatic interruption experiment in a plasma switch was performed in 1940 in a mercury thyratron by H. Fetz.[27] Fetz found that at a suitably low gas pressure, a negative bias of the grid caused a constriction of the current channel through each grid aperture and an increase of the forward discharge voltage drop across the switch. When the forward drop increased beyond the power supply voltage, the thyratron discharge was extinguished and the current was interrupted. In 1954 E.O. Johnson[2] developed a variant of the thyratron which he called a tacitron. This device operated in a unique discharge mode where ion generation occurred solely in the grid-to-anode gap. This mode purportedly allowed operation with reduced noise and with current interruption capability through application of negative potentials to the grid. Recently, A.E. Hill[28] also reported on a grid-controlled thyratron-type switch where the hot cathode is replaced by the wire-anode cold-cathode-discharge technique[20,21] described earlier.

Despite the demonstration of current interruption in these experiments, opening-switch capability has never exceeded a few tens of amperes and development has never proceeded to commercialization. However, the thyratron industry in the USSR apparently recognized the utility of an opening plasma switch and developed their own version[3] of the tacitron. Although the Russian literature does not make clear the details of the discharge mode in their switch, they have adopted the name "Tacitron", and have developed commercial devices capable of switching 300 A at 12 kV with 100-kHz PRF.

Basic Principles

The CROSSATRON Modulator Switch is based upon a crossed-field discharge in a four-element, coaxial system consisting of a cold cathode, two grids, and an anode, as shown in Fig. 10. In a manner analogous to thyratron operation, charges for conduction are generated by a plasma discharge near the cathode. However, in the CMS the plasma is produced by a crossed-field cold-cathode-discharge technique (using crossed electric and magnetic fields as in a XFT) in a gap located between the source grid (which serves as the anode for the local crossed-field discharge) and the cathode. The gap is magnetized with a cusp field supplied by permanent magnets attached to the outside of the switch. This arrangement eliminates the need for cathode-heater power and also permits instant-start operation. The switch is closed by pulsing the second control-grid electrode above the plasma potential to allow conduction of charges to the anode. The anode voltage then falls to the 200- to 1000-V forward-drop level and plasma fills the switch volume.

Once a plasma discharge is initiated, grid control is no longer possible for most gaseous-discharge devices. In a thyratron, for example, if current interruption is attempted by returning the control grid to the cathode potential or below, plasma will continue to flow through the grid to maintain conduction. However, in the CMS system grid control and current interruption can be maintained for cathode current densities up to 10 A/cm^2. This unique feature of the CMS is enabled by five key elements:

1. <u>Grid Structure</u>: High transparency grids (80 percent) with small, but practical sized, apertures (0.32 mm dia.) are used which are produced by chemical-etch techniques.

2. <u>Control Grid Position</u>: The control grid is located as close to the anode as allowed by vacuum breakdown considerations to insure high-speed switching.

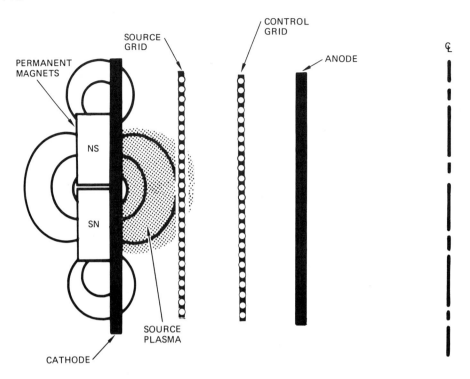

Fig. 10. CROSSATRON Modulator Switch configuration.

3. <u>Localized Ionization Source</u>: Using a highly localized cusp magnetic field near the cathode, ionization occurs primarily in the cathode-to-source-grid gap, allowing high cathode current density with relatively low plasma density near the control grid.
4. <u>Low Pressure</u>: Low helium or hydrogen gas pressure (50 mTorr), enabled by the use of the crossed-field discharge, is used to maintain a low plasma density near the control grid and to avoid Paschen breakdown.
5. <u>Controlled Space-Charge Neutralization</u>: High current densities are controlled with a low voltage drop by providing ion-space-charge neutralization of the electron current in the control-grid-to-anode gap.

With the ionization source being highly localized near the cathode and the control grid being positioned near the anode, the ion density in the vicinity of the control grid is low (relative to the cathode). As shown below, this low-density ion flux allows current interruption by applying a negative potential, relative to the plasma, to a grid having small yet finite-sized (0.3 to 1-mm diameter) apertures. Through application of a negative potential, an ion sheath is created around the grid which permits plasma cut-off to the anode region provided the sheath size is larger than the grid-aperture radius. Upon plasma cut-off, switch current is interrupted as the remaining plasma in the control-grid-to-anode gap decays. Low pressure operation insures that ionization cannot sustain the plasma in the narrow, isolated control-grid-to-anode gap where the magnetic field is very low ($< 10^{-3}$ T). Figure 11 demonstrates the generation of fast, square-wave pulses by using this grid-control technique.

Operating Mechanisms

Operation of the CMS by electrostatic control of grids is shown schematically in Fig. 12. As discussed above, charges for conduction are provided by a low-pressure gas discharge in the source section of the switch. The source plasma is generated - Fig. 12 (a) - by pulsing the potential of the source grid (SG) electrode to > 200 V for a few microseconds to establish a crossed-field discharge. When equilibrium is reached, the SG potential drops to the glow-discharge level, ~ 200 V above the cathode (C) potential. With the control grid (CG) remaining at cathode potential, the switch remains open and the full anode (A) voltage appears across the CG-to-A vacuum gap. The switch can now be

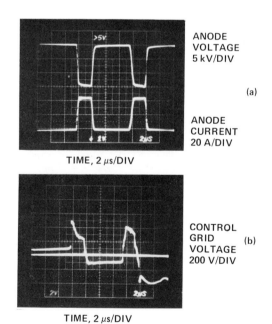

Fig. 11. Anode current and voltage waveforms (a) and control-grid-drive waveforms (b) for dual-pulse modulator operation at 45 A and 15 kV.

closed (the anode switched ON) by releasing the CG potential, or by pulsing it momentarily above the 200-V plasma potential. As plasma streams through the CG, electrons are collected by the anode, the switch conducts, and the anode voltage falls to the 200-V level, as shown in Fig. 12 (b). In order to open the device (i.e. switch the anode OFF, Fig. 12 (c)), the CG is returned to the cathode potential or below in a hard tube fashion.

This last operation is not usually successful in most other plasma switches. Depending upon the size of the grid apertures, the potential of the grid relative to the plasma, and the local ion density, plasma may continue to stream through the CG to the anode region to maintain conduction. Also, even if plasma is cut off by

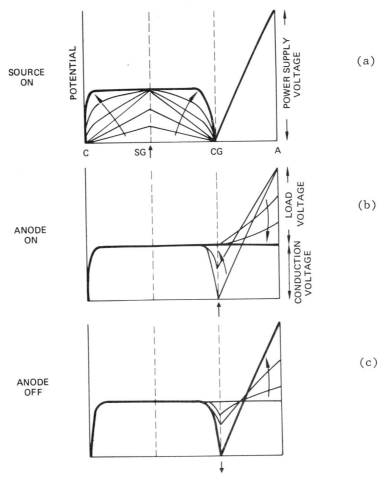

Fig. 12. CROSSATRON Modulator Switch operation through electrostatic control of grids.

the grid, conduction may persist if the gas pressure is high enough to sustain ionization in the CG-to-A gap. Thus, successful current interruption in a plasma switch depends upon low gas pressure, a controlled magnetic-field profile, and upon the physics of the grid-plasma interaction, the details of which we now focus our attention.

Beginning with Fig. 13 (a), we observe that when the CG voltage is raised to the plasma potential, plasma from the source section is drawn through the grid to fill the CG-to-A-gap (Fig. 13 (b)) and enables switch conduction. Under these conditions, current can be drawn with low potential drop through the grid apertures. If the grid voltage is now driven below the plasma potential (Fig. 13 (c)), the grid will begin to draw ion current and an ion-space-charge-limited sheath will appear between the plasma and the grid. The amount of ion current drawn depends upon the plasma density and temperature; and the size of the sheath (Δx) is determined by the ion current density (J) and the voltage difference (V) between the CG and plasma. The functional relationship between J, Δx, and V is given by the Child-Langmuir sheath theory which was discussed above and is summarized by Eq. (3). If, as shown in Fig. 13(d), the ion current is sufficiently low and the voltage is sufficiently high that the sheath dimension expands beyond the radius of the grid aperture, then plasma cut-off is achieved and ions can no longer diffuse to the right of the grid into the anode region without being repelled by the grid potential. As the now-isolated plasma in the CG-to-A gap begins to dissipate (e.g. by erosion), charges for conduction are lost and the anode current is interrupted, provided that the gas pressure is low

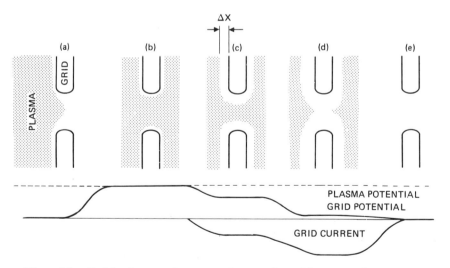

Fig. 13. Grid-plasma interaction and grid-control process.

enough that ionization is not sustained in the gap. Immediately following interruption (Fig. 13 (e)), the grid region is nearly free of plasma.

In thyratrons and other higher-pressure devices (ignitrons and spark gaps), the conditions described in the preceding paragraphs are not satisfied and plasma cut-off is not achieved due to high plasma densities and the associated very small sheaths. Consequently, current interruption by electrostatic grid control is not possible. In the CMS, however, low-pressure operation is made possible by the crossed-field discharge. Electron trapping in the cusp magnetic field leads to rapid (but localized), high density plasma production near the cathode of the C-to-SG gap at low pressure. Furthermore, as a consequence of the localization of plasma near the cathode by the cusp magnetic field, the plasma density falls sharply toward the anode and leads to large sheaths near the CG.

Since the ion current density is low near the CG and anode, high-current interruption can be maintained in the CMS with reasonably-sized control grid apertures. This capability is illustrated in Fig. 14 which shows the results of experiments performed to determine the scaling of maximum-interruptible switch current with control-grid aperture size. The data points indicate that currents up to 250 A (current densities up to 7 A/cm^2) are interrupted in a small device with a grid having 0.32-mm-diameter apertures. Fig. 14 also shows that the maximum interruptible current density increases as the gas pressure is reduced. This scaling is also anticipated since lower gas pressure leads to lower plasma density and larger ion sheaths.

Once plasma cut-off at the control grid is achieved, the switch current is interrupted on a time scale determined by the ion transit time across the CG-to-A gap. If the gap size is larger than an ion-sheath thickness, then ions are lost at the ambipolar rate which leads to an opening time given by Eq. (1). If the ion density is very low or the applied negative voltage to the grid is very high, such that the ion sheath becomes larger than the gap size, then ions can be accelerated out of the gap at super-ambipolar speeds.

CMS Construction and Performance

CROSSATRON Modulator Switches are constructed in a manner which is similar to the XFT except for the two grid electrodes which are mounted on small feed-through bushings in the space between the cathode and anode. A schematic diagram of a 15-cm-diameter CMS is shown in Fig. 15. Denoted XTRON-3, this switch is actually rated beyond the capability of most power-tetrode devices at 40 kV and 500 A. The XTRON-3 device is constructed with an all

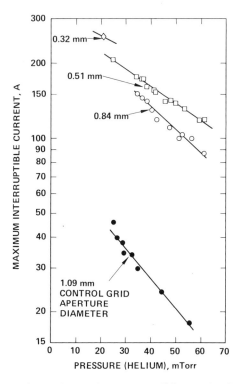

Fig. 14. Scaling of maximum-interruptible switch current with gas pressure and control-grid aperture size.

metal-ceramic, vacuum-sealed envelope and employs a small titanium-hydride internal gas-pressure controller. A photograph of this "productized-prototype" switch is shown in Fig. 16. Oscilloscope waveforms which demonstrate XTRON-3 performance in an IES circuit at 450 A and 45 kV (with 1-μs opening time) are shown in Fig. 17.

Hughes has also developed a high-average-power CMS for command-charging modulator service. Many pulsed-power systems employ series-resonant, command-charging circuits to efficiently control the charging of capacitor banks and PFNs. In recognition of its cold-cathode, low-forward-drop, high-gain, and current-interruption operating features, the CMS maintains a unique synergism with the performance required by command-charging circuits for pulsed-power systems such as high-energy lasers, particle accelerators, and high-power microwave devices. The 25-cm (cathode diameter) CMS is called XTRON-4 and was designed for 100-kV, 2.5-kA, and 5-MW (average power) ratings. Since it must operate in series with the output of a 100-kV power supply, it was integrated into the corona-shielded, high-voltage floating deck shown in Fig. 18.

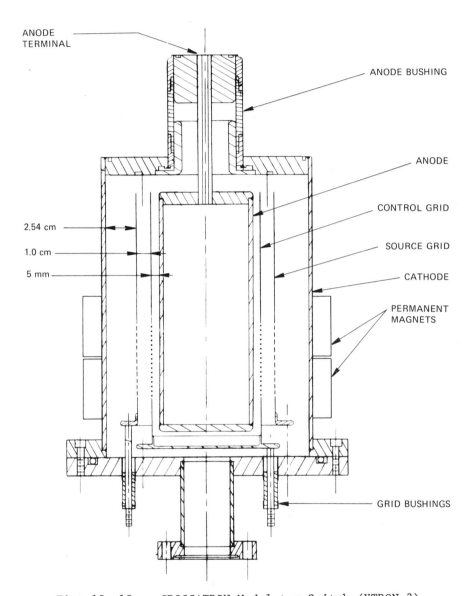

Fig. 15. 15-cm CROSSATRON Modulator Switch (XTRON-3).

LOW-PRESSURE PLASMA OPENING SWITCHES 119

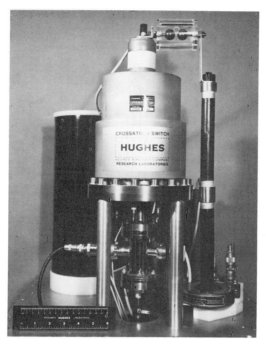

Fig. 16. Photograph of the "Productized-Prototype" 15-cm CROSSA-TRON Modulator Switch (XTRON-3).

In order to simulate operation of XTRON-4 in a high-power command-charging circuit, it was tested in a single-pulse LC circuit consisting of a 5-μF capacitor bank and a 2.3-H inductor. The capacitor bank is charged to 80 kV and then discharged by XTRON-4 into the inductor. Near the end of the 10-ms, half-sinusoid current pulse, the CROSSATRON switch is opened and the series current is interrupted. A varistor stack limits the inductive voltage kick and absorbs the energy remaining in the inductor. Current and voltage waveforms that demonstrate switch operation in this circuit are shown in Fig. 19. Switch conduction is controlled with the fiber-optic gate pulse shown near the bottom of the oscilloscope photograph. When XTRON-4 is gated on, the voltage across the inductor rises to the 80-kV capacitor-bank voltage and the switch current begins to rise. A small increment of instantaneous current is drawn by the varistor. The series-resonant current peaks at 110 A; after 8.5 ms of conduction, the switch is gated off. With the capacitor bank reverse-charged to -46 kV, the inductor voltage suddenly rises until the varistor limits the voltage at -120 kV. At this point the CROSSATRON switch successfully holds off the 74-kV open-circuit voltage difference without reinitiating conduction.

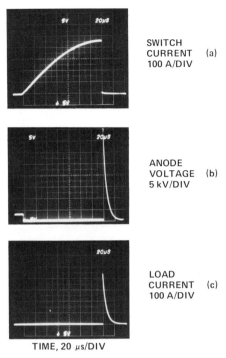

Fig. 17. Oscilloscope traces of switch-current, anode-voltage, and load-current waveforms for XTRON-3 operation in an IES circuit.

The XTRON-4 was also tested in a multipulse, multimegawatt command-charging circuit at AVCO Everett Research Laboratories. As shown in Fig. 20, it successfully demonstrated the principle of command-charge/command-stop operation at average power levels up to 1 MW. Pulse-length control and voltage-regulation capability were demonstrated by interrupting the charge current on the falling edge of the pulse at the 20-A current level with 52-kV power-supply voltage and 100-Hz PRF. At lower frequency and current, operation was also demonstrated at 62-kV open-circuit voltage and 12-ms pulse lengths with 66-s-long bursts.

ULTIMATE CAPABILITIES

Summary

The ultimate capabilities of low-pressure plasma opening switches are summarized in Table 1 (parameters are not necessarily obtained simultaneously) where for each switch parameter the demonstrated and ultimate performance values are listed, together with the limiting physical mechanism(s). The first six of the nine

Fig. 18. High-voltage-isolated XTRON-4 switch system in a single-pulse, LC test circuit. The switch is on the left and the high-voltage inductor is on the right.

parameters in the table have already been discussed. For example, the maximum open-circuit voltage (V_{open}) per gap, minimum forward-voltage drop (V_{closed}), and maximum cathode current density (J_c), were examined in detail above. We have discussed how the forward-voltage drop depends upon cathode surface physics (secondary electron yield), but we should also mention here that it may well be possible to obtain a significantly lower forward drop, ~ 30 V, for special applications (such as those that require very high current conduction with low tube dissipation). This may be accomplished by using a cathode with an appropriate coating which reduces the work function and increases the secondary yield.

We also examined the mechanisms which determine the opening time and found two regimes: the ambipolar regime where the interruption time scales as Eq. (1), and the super-ambipolar regime where electric fields are applied in order to accelerate ions out

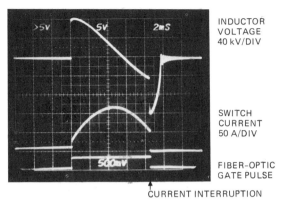

Fig. 19. Inductor-voltage, switch-current and fiber-optic gate-pulse waveforms that demonstrate XTRON-4 operation in a single-pulse LC circuit.

of the gap at super-ambipolar speeds. Using this technique, interruption times as fast as 50 ns were obtained at a low current density < 1 A/cm^2. Opening times as fast as 20 ns may ultimately be possible by using a higher-speed control-grid pulser circuit capable of driving the grid $\geq$ 1 kV below the cathode potential at interruption. Our optimism in this ultimate performance is based on demonstrated closing times which are as short as 20 ns when the control grid is driven $\geq$ 1 kV above cathode potential with a high-speed pulser. In this reciprocal operation, ions are accelerated into the gap in order to attain fast closing.

The ultimate pulse repetition frequency (PRF) is limited by the opening time and by control-grid heating. Obviously the switch cannot be re-closed faster than it is opened, so if the switch is opened faster than 1 μs (as demonstrated by the CMS), an ultimate PRF of 1 MHz can be obtained. An effective PRF of 250 kHz has been demonstrated for a two-pulse burst. A PRF of 10-kHz CW and 16 kHz in a long burst has also been demonstrated using the same CMS in a line-type-modulator circuit where the switch was required to recover quickly, but was not required to interrupt high current directly. Control-grid power dissipation is also a consideration in determining the ultimate PRF. Depending upon the control-grid structure used for the particular application and upon the switching speed and current desired, the grid may reach temperatures where deformation or thermionic emission would limit the PRF to values below 1 MHz.

The last three switch parameters in Table 1 have not been examined in detail previously. Therefore, separate discussions of maximum current and useable life are now presented.

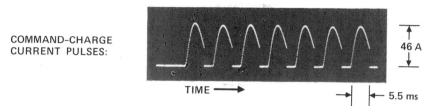

Fig. 20. Command-charge/command-stop current pulses which were modulated by XTRON-4.

Table 1. Ultimate Capabilities of Low-Pressure Plasma Opening Switches

SWITCH PARAMETER	DEMONSTRATED PERFORMANCE	ULTIMATE PERFORMANCE	LIMITING MECHANISM
V_{OPEN}	120 kV	200 kV	VACUUM AND PASCHEN BREAKDOWN
V_{CLOSED}	200- 1000 V	30 V	SURFACE PHYSICS
J_C	10 A/cm^2	10 A/cm^2	GAT
$\tau_{opening}$	50 ns	20 ns	ION ACCELERATION
$\tau_{closing}$	20 ns	20 ns	ION ACCELERATION
PRF	16 kHz	1MHz	$\tau_{opening}$ GRID HEATING
I_{TOTAL}	10 KA	50 KA	OVERALL SIZE, SELF-FIELD EFFECTS?
$I_{AVE.}$	40 A	?	CATHODE HEATING
LIFE	1.5 X 10^9 SHOTS	>10^{11} SHOTS	ION SPUTTERING

Maximum Current

Although the recently developed CROSSATRON Modulator Switch has been scaled to a total current level of only 500 A, a considerable amount of effort has been devoted to scaling the total interruptible current capability of crossed-field tubes to ~ 10 kA. The early work on XFT scaling was directed at increasing the current density, J, to keep the tube size to a minimum. This work led toward high-purity (i.e., inclusion free), refractory-metal cathode materials and ultra-clean assembly and vacuum techniques. Reliable operation (meaning operation with an arc probability at interruption of less than 1 percent) was demonstrated with these small tubes for cathode current densities of up to 5 - 10 A/cm^2.

More recently, it was decided to pursue larger tubes and lower current densities because the reliability increase at lower densities more than offsets the difficulties associated with larger tube size. XFTs with cathode surface areas of 50, 100, and 500 cm^2 are found to have nominal interruptible currents (at voltages of 50 to 70 kV) of 250, 500, and 2700 A, respectively. However, devices with larger surface areas do not exhibit this area scaling (i.e., $J \leq 5$ A/cm^2), and J falls to as low as 1.2 A/cm^2 with a cathode surface area of 6000 cm^2.

Empirically, it has been found that the maximum, reliably interruptible current, I_m, scales as the ratio of the cathode radius to the gap spacing. This scaling is shown in Fig. 21 for crossed-field tubes having cathode areas up to 8000 cm^2 with gap spacings in the range $1.0 < d < 1.9$ cm. Experimentally, d is determined by the voltage holdoff desired, and it is not practical to reduce it much below this range. This leaves the radius as the critical factor. The axial length of the electrodes is not found to be important. In fact, when the active length of an XFT is varied by varying the length of the magnetic-field coil, no significant change in I_m is observed. Field coils energizing only a few centimeters of tube length are nearly as effective as coils tens of centimeters long. Therefore, we take the empirical scaling of the interruptible current with tube dimensions to be given by the law:

$$I_m \sim 540 \ r/d \ . \tag{8}$$

There are several plausible explanations for this empirical current scaling. One is the fact that it becomes increasingly difficult to condition the electrodes to operate at high current density without arcing as the electrode area increases. Another is the simple, statistical increase in the arc rate which is expected as the area scales up. A third explanation is based upon the self-magnetic field generated by the switch current. A current, I, flowing through the cylindrical anode (the inner electrode) generates a magnetic field B oriented in the azimuthal direction:

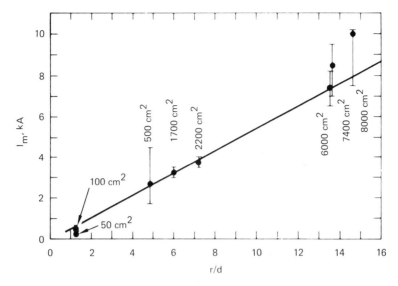

Fig. 21. Empirical scaling of XFT maximum, reliably interruptible current.

$$B_\theta = \frac{\mu_o I}{2\pi r}. \qquad (9)$$

Substituting I_m from Eq. (8) and using d = 1.5 cm, we calculate $B_\theta = 7 \times 10^{-3}$ T (70 G). This is typically equal to the value of the applied axial magnetic field, B_z, required to sustain conduction. The axial field produces a circumferential ($\bar{E} \times \bar{B}$) drift of electrons around the anode while the self-generated field due to the current flowing in the anode produces an axial drift (in the direction of the anode current).

The fact that when I_m is flowing, arcing occurs during interruption leads to the following hypothesis: The B_θ field adds to the B_z field, thus increasing the magnitude of the magnetic field and changing its direction. During conduction, this has no deleterious effects. When B_z is reduced, however, B_θ is sufficient in magnitude to trap electrons and maintain the discharge. But since the $\bar{E} \times \bar{B}$ drift direction is now predominantly axial, the plasma will be forced into a new location. The original uniform current density is distorted by this shift in the plasma position, especially at the edge of the original discharge region. If the current density fluctuation is large, an arc is likely to result. Experiments performed with an externally applied axial current (to vary B_θ independent of switch current)[7,29] confirm that the axial current density distribution becomes unstable near $B_\theta = 7 \times 10^{-3}$T.

In the CROSSATRON Modulator Switch, the current is interrupted without changing the applied field near the cathode. The CMS may therefore have a higher maximum interruptible current when it is scaled to large sizes. We have therefore estimated the maximum total current per device to be about 50 kA in Table 1.

Because of the relatively large cathode sheath drop shown in Fig. 3, most of the power dissipated in a tube is absorbed by the cathode. Since the cathode is the outer-most electrode it is easy to cool the cathode surface and therefore to operate the low-pressure switch at high average current. This is particularly true during operation at low PRF with long pulses where the switching rate is low and the grid-drive power (in the CMS case) is low. The XTRON-4 device, for example, employs a water-cooled cathode and has operated at 19-A average current with 8.5-ms pulse lengths and 100-Hz PRF. However, at high PRF (> 1 kHz) the power absorbed by the grids and anode during switching may become significant and may therefore limit the average current unless they are actively cooled also. Thus, ultimate average current for any given tube depends upon the application (pulse length, PRF, rise and fall times), so it is difficult to estimate this particular ultimate-performance parameter. Qualitatively, we can say that I_{ave} is very high. For example, a 1-MW average-power, magnetically modulated, grid-triggered, hybrid XFT[30] has demonstrated 40-A average-current operation (50 kV, 50 kA, 10-μs pulse at 80 Hz) in a line-type modulator circuit.

Lifetime

As mentioned earlier, the life of a low-pressure, cold-cathode plasma switch is expected to be long since there are no hot filaments or oxide-coated hot cathodes to burn-out or poison. The internal structure of the switch is degraded only by ion sputtering, and sputter lifetimes are rather long. Although a sputter life of only 100 Coul/cm^2 was observed in early crossed-field tubes[4] (limited by flaking of sputtered metal on the anode), recent life tests of the CROSSATRON Modulator Switch have recorded more than 2400 Coul/cm^2 of charge passed through the switch without a single failure or any operational degradation. This corresponds to more than 1.5 billion square-wave modulator pulses at 25 kV and 40 A. The significant increase in sputter lifetime is believed to be due to improved materials and vacuum conditions, and the shielding of the cathode from the high-voltage anode electrode by the grids.

The ultimate sputter lifetime of a CMS device can be estimated using the following arguments. Ions are collected at the cathode wall during conduction and by the control grid during interruption. However, since the cathode is usually a thick tube while the control grid is significantly more thin, ion sputtering is expected to cause the control grid to fail first. Also, the cathode is sputtered by only low-energy ions (~ 200 to 500 eV) while the control grid is sputtered by ions with energy up to near the full open-circuit anode voltage during interruption. The control grid is ultimately expected to fail by ion-sputter erosion of the grid-aperture web to the point where small apertures become large apertures and the current interruption capability is suddenly degraded (see Fig. 14). The number of switch operations, N_s, (shots) obtained before this failure occurs depends upon the ion current density; J_i, in the switch; the opening time, τ_o; the grid thickness, ℓ_g; and the anode voltage (through the sputter-erosion rate, R); according to:

$$N_s = \frac{\ell_g e \rho (1-T)}{J_i \tau_o R} \tag{10}$$

where e is the electronic charge, ρ is the atom density in the grid, and T is the grid transparency.

Conservative estimates of the control-grid sputter lifetime have been made based upon sputter-erosion rates measured in high vacuum. These estimates are conservative since erosion rates on gas-covered electrodes typical of a 50 mTorr vacuum environment (characteristic of the CMS) are usually much lower. Data for hydrogen ions sputtering molybdenum for this case is not available. Using the published erosion rate of 10^{-2} sputtered molybdenum atoms per incident hydrogen ion in high vacuum, we find a typical 0.13-mm-thick, molybdenum control grid to have a sputter life of 10^{11} shots, assuming an ion current density of 0.5 A/cm^2, an opening time of 75 ns, and an anode voltage of 25 kV. This conservatively estimated sputter lifetime is equivalent to 28,000 modulator operating hours at 1-kHz PRF. Such a lifetime is quite satisfactory for advanced modulator applications. In higher current applications (such as in IES circuits) the current density will be ten-times higher, but a larger ℓ_g may be used. Sputter lifetimes on the order of 10^{10} shots may be expected in this case.

Other possible life-limiting problems are the deposition of sputtered metal on insulator surfaces, metal blister and flake formation (which may degrade voltage hold-off capability), and depletion of the gas reservoir. Suitable baffling of the insulator regions has been found to prevent metal deposits. Also, gas reservoir depletion is not expected to be a severe problem since CMS devices use thyratron-type (titanium-hydride) hydrogen reservoirs (operating at low temperatures and low pressures) which generally provide a 500- to 10,000-hour life. Little is known about metal blister and flake formation in the switch environment. Future CMS development will prove whether this problem is more severe than control-grid sputter damage.

REFERENCES

1. R.E.B. Makinson, J.M. Sommerville, K.R. Makinson, and P. Thoneman, Magnetically-Controlled Gas Discharge Tubes, J. Appl. Phys., 17:567 (1946). See also, G. Boucher and D. Doehler, "The Artatron-A Switch Tube for High Voltage and High Current," Proc. 3rd Symp. on Engineering Problems in Thermonuclear Research, Munich, June 22-24, 1964.
2. E.O. Johnson, J. Olmstead, and W.M. Webster, The Tacitron, A Low Noise Thyratron Capable of Current Interruption by Grid Action, Proc. of the I.R.E., 25:1350 (1954).
3. V.D. Dvornikov, S.T. Latvshkin, V.A. Krestov, L.M. Tikhomirov, and L.P. Yudin, Powerful Tacitrons and Some of Their Characteristics in a Nanosecond Range, Pribory I Tekhnika Experimenta, 4:108 (1972). See also, T.R. Burkes, J.P. Craig, M. Gundersen, E.E. Kunhardt and W.P. Portnoy, "Switches for Directed Energy Weapons: An Assessment of Switching Technology In the USSR," Final Report Under Contract No. DAAK21-82-C-0111 To Army Foreign Science and Technology Center, Charlottesville, VA, May 18, 1984.
4. M.A. Lutz and G.A. Hoffman, The Gamitron - A High-Power Crossed-Field Switch Tube for HVDC Interruption, IEEE Trans. Plasma Science, PS-2:11 (1974).
5. R.J. Harvey and M.A. Lutz, High Power ON-OFF Switching with Crossed-Field Tubes, IEEE Trans. Plasma Sci. PS-4:210 (1976).
6. R.J. Harvey, M.A. Lutz, and H.E. Gallagher, Current Interruption at Powers Up to 1 GW With Crossed-Field Tubes, IEEE Trans. Plasma Sci., PS-6:248 (1978).
7. R.J. Harvey, The CROSSATRON: A Cold Cathode Discharge Device With Grid Control, Conference Record of the Fourteenth IEEE Power Modulator Symposium, 77-79, June 1980.
8. R.W. Schumacher and R. J. Harvey, CROSSATRON Modulator Switch, Conference Record of the Sixteenth IEEE Power Modulator Symposium, 139-151, June 1984.
9. J.H. Ingold, Chapter 2, in "Gaseous Electronics," M.N. Hirsh and H.J. Oskam, eds. Academic Press, New York, (1978).
10. M.J. Schonhuber, Breakdown of Gases Below Paschen Minimum: Basic Design Data of High-Voltage Equipment, IEEE Trans. Power Apparatus and Systems, PAS-88:100 (1969).
11. F.F. Chen, "Introduction to Plasma Physics", Plenum Press, New York (1974), p. 139-140.
12. R.V. Latham, "High Voltage Vacuum Insulation: The Physical Basis," Academic Press, New York (1981), p. 13-46.
13. R.L. Jepsen, Magnetically Confined Cold-Cathode Gas Discharges at Low Pressures, Jour. Appl. Phys., 32:2619 (1961).
14. S.C. Brown, "Basic Data of Plasma Physics," John Wiley & Sons, New York (1959), p. 227-238.

15. D. Bohm and E.P. Gross, The Theory of Plasma Oscillations, Phys. Rev., 75:1851 and 1869 (1949).
16. J.H. Holliday and G.G. Isaacs, Arc Initiation at Metal Surfaces in A Hydrogen Penning Discharge, Brit. Jour. Appl. Phys., 17:1575 (1966).
17. M.A. Lutz, The Glow-to-Arc Transition - A Critical Review, IEEE Trans. Plasma Science, PS-2:1-10 (1974).
18. J. Millman, "Vacuum-Tube and Semiconductor Electronics," McGraw-Hill, New York (1965), pp. 22-23.
19. M.A. Lutz, R.J. Harvey and H. Alting-Mees, Feasibility of a High-Average-Power Crossed-Field Closing Switch, IEEE Trans. Plasma Sci., PS-9:118 (1976).
20. J.R. Bayless and R.J. Harvey, "Continuous Ionization Injector for Low Pressure Gas Discharge Devices," U.S. Patent No. 3,949,260 (1976).
21. G.W. McClure, Low Pressure Glow Discharge, App. Phys. Lett., 2:233 (1963).
22. G.A. Hoffman, G.L. LaBarbera, N.E. Reed, and L.A. Shillong, A High Speed HVDC Circuit Breaker With Crossed-Field Tubes, IEEE Trans. Power App. and Sys., PAS-95:1182 (1976).
23. G.A. Hoffman, Technical Manual For A 40-A, 120-kV Crossed-Field Interrupter For Neutral Beam Protection. Prepared by Hughes Research Laboratories for the Culham Laboratory, Abingdon, U.K., 1978.
24. H.J. King, H.E. Gallagher and W. Knauer, 145-kV Current Limiting Device - Design, Construction, and Factory Test, IEEE Trans. Power App. and Sys. PAS-99:911 (1980).
25. R.J. Harvey, "Cold Cathode Discharge Device with Grid Control," U.S. Patent No. 4,247,804 (1981).
26. R.W. Schumacher and R.J. Harvey, "Modulator Switch with Low-Voltage Control," U.S. Patent No. 4, 596, 945 (1986).
27. H. Fetz, Uber die Beeinflussung eines Quecksilbervakuumbogens mit einem Steuergitter in Plasma, Ann. der. Phys., 37:1 (1940).
28. A.E. Hill, A Novel Grid Controlled Switch Incorporating Isolated Neutral Plasma and Free Electron Flow Regions, IEEE Trans. Plasma Science, PS-10:302 (1982).
29. R.J. Harvey, Paramagnetic Crossed-Field Discharges, Proceedings of the First International Conference on Plasma Science, Knoxville, Tennessee, p. 15 (May 1974).
30. R.J. Harvey, Operating Characteristics of the Crossed-Field Closing Switch, IEEE Trans. Elect. Dev., ED-26:1472 (1979).

RECOVERY OF DIELECTRIC STRENGTH IN GASES

R.N. DeWitt

Naval Surface Weapons Center
Dahlgren, VA 22448

INTRODUCTION

   Normally, when referring to the recovery of a switch, one refers to the recovery of the dielectric strength of the insulating medium; that is, the maximum voltage the medium can hold off without breakdown. After sufficiently long time the medium can return to the predischarge condition and is said to be fully recovered, assuming complete reversibility in the medium and electrode characteristics. While switch media may be gaseous, liquid, solid, or vacuum, only the gaseous case will be considered in this chapter. Predischarge (and thus fully recovered) dielectric strengths have been studied for many years with the aim of establishing criteria that quantify the dielectric strength of these insulating media or the mechanisms and processes leading to the breakdowns of the media. The classical treatments can be found in Ref's. 1-6, for example, while recent research is found not only in physics and engineering journals but also in edited proceedings, such as Ref's. 7-10. A collection of important papers with comments can be found in Ref. 11. Predischarge and breakdown studies are most important in problems of insulation and single shot switching. However, it it unclear how much of this information applies to switching at high repetition rates. This chapter will focus on the main issues and indicate the important uncertainties and over-simplifications rather than provide an extensive review of the literature.

   A fully recovered gas will withstand an applied voltage (secondary voltage) equal to the predischarge voltage (the primary voltage). If it cannot, then the ratio of the maximum secondary voltage to the primary voltage is referred to as the percent of recovery. The essential point is that the percent recovery is determined through an applied voltage. The two main questions are:
1) what are the electrical, thermal and fluid conditions of the

medium at any time after the primary discharge, and 2) what are the breakdown criteria under these conditions?

The purpose of a switch in a circuit is to control the energy. The closing switch remains open under high voltage until predetermined conditions allow it to conduct. The switch then transfers stored energy in a characteristic time, and if it is operating in a repetitive mode, it must recover to function again in a similar manner. The opening switch on the other hand must stay closed for a given time while energy is being stored in a magnetic field and then must open against high voltage in some characteristic time. Again, if in repetitive operation, it must recover to function again in the same manner. The energy budget during the discharge and postdischarge period is depicted in Fig. 1.

Here, the postdischarge is distinguished from recovery because, as mentioned previously, recovery implies reestablishment of the dielectric strength and thus addresses the postdischarge state with an electric (and possibly magnetic) field applied. Being a state with high gas temperature, partially ionized species, and interactions under an imposed field, this distinctive state then challenges the applicability of the Paschen Law for Townsend avalanches and Raether's criterion for streamer formation. It can also be seen that in the early stages of recovery there are fluid-dynamic, electrodynamic and thermodynamic processes occurring.

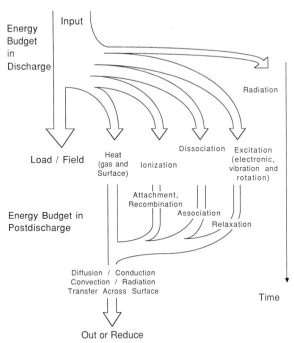

Fig. 1. Energy budgets during discharge and recovery.

While gas density may be the dominant parameter at late recovery times, it is unclear what dominates the early recovery period, which is important for high repetition rate switching. It would be helpful to determine intrinsic time constants for the multitude of processes involved, then the given repetition rate would establish the appropriate set of processes for modeling or diagnostic investigations.

INITIAL BREAKDOWN

If a voltage is applied to electrodes separated by a distance d, with the gap containing a gas, then for a sufficiently small voltage, electrons move through the gas with an average velocity (drift velocity u) determined by their mobilities, μ, in the given gas and satisfy the equation

$$u = \mu E , \qquad (1)$$

where E is the electric field intensity and is taken, for a uniform field, as $E = V/d$, where V is the applied voltage. If the voltage is increased to a level where the electrons gain enough energy from the field between collisions with the gas molecules to cause ionizing collisions, then a multiplication of electrons can occur and the electron density, n_e, is governed by the continuity equation

$$\partial_t n_e + \mathbf{v} \cdot n_e \mathbf{u} = \nu_I n_e - \alpha_R n_e^2 - \nu_a n_e + \ldots , \qquad (2)$$

where **u** is the average velocity, ν_I the electron collision frequency for ionization, α_R the electron-ion recombination coefficient, and ν_a the electron-neutral attachment frequency. Other source and loss terms could be added. For the steady state, one dimensional, constant drift velocity, ionization only case,

$$dn_e = \frac{\nu_I}{u} n_e \, dx = \alpha_T n_e dx , \qquad (3)$$

where $\alpha_T = \nu_I/u$ is the Townsend ionization coefficient and represents the number of ionization collisions per unit length.*

With the current density $j = enu$, the current grows exponentially At sufficiently high fields the ions produced in the electron multiplication process can be accelerated to the cathode and cause

---

* These derivations and other aspects of gas breakdown theory are described in more detail in the chapter by E. Kunhardt in the next volume (Gas Discharge Opening Switches) of this series.

secondary electrons to be emitted and their numbers amplified. The runaway condition in this process is the Paschen Law, shown schematically in Fig. 2.

The voltage necessary for a self-sustained discharge, the breakdown voltage, is a function of the product of the pressure and gap spacing. According to this law, the breakdown voltage of a gas is determined solely by the minimum density within the gas. All of this is developed for electrons flowing through a gas of neutral particles in a constant, uniform electric field. Their ionizing collisions lead to an avalanche multiplication process (Townsend avalanche). If electrons can be lost in the process by being attached to neutral molecules, Eq. (3) is modified by the third term on the RHS of Eq. (2) giving

$$dn_e = \alpha_T n_e dx - \eta n_e dx = (\alpha_T - \eta) n_e dx = \bar{\alpha} n_e dx ,\qquad (4)$$

where $\eta = \nu_a/u$ is the attachment coefficient and $\bar{\alpha} = \alpha_T - \eta$ is called the net ionization coefficient. Thus, if $\eta > \alpha_T$, the avalanche will quench.

The generally accepted breakdown criterion (Paschen's Law) is based on the current density at the anode becoming infinite in a runaway process for a given gas density and electrode spacing. It is possible at high pressure that the space-charge fields at the tip of the Townsend avalanche exceeds the applied field in the gap before reaching the anode. At this point a so-called streamer rapidly develops (in the order of 10^{-8} s) and breakdown follows immediately. If the number of electrons at an avalanche length x is $e^{\alpha x}$ and are considered to be in a spherical volume of radius r, then the space-charge field, E_r, at the surface of the avalanche tip is given by

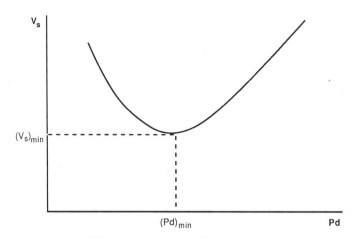

Fig. 2 The Paschen Curve

$$E_r = \frac{q_e \left[e^{\alpha_T x}\right]}{4\pi\epsilon r^2} , \qquad (5)$$

where q_e is the electron charge. If the radius r is the result of electrons diffusing transversally as the avalanche develops, then the radius is proportional to the square root of time in going the distance x so that

$$r \approx \sqrt{4Dt} , \qquad (6)$$

where D is the diffusion coefficient. With $t = x/u = x/\mu E$, the space-charge field becomes

$$E_r = \frac{q_e \left[e^{\alpha_T x}\right]}{4\pi\epsilon 4xD/\mu} E . \qquad (7)$$

Using the Einstein diffusion relation

$$\frac{D}{\mu} = \frac{kT_e}{q_e} , \qquad (8)$$

where T_e is the electron temperature and k is Boltzmann's constant, the criterion (Raether) for streamer formation ($E_r \approx E$) becomes

$$x_s = \frac{\mu q_e \left[e^{\alpha_T x_s}\right]}{16\pi\epsilon D} = \frac{q_e^2 \left[e^{\alpha_T x_s}\right]}{16\pi\epsilon \, kT_e} . \qquad (9)$$

The Townsend and Raether theories apply to limited electron production in cold gases by a uniform electric field. Characteristic properties of the gas (α_T, η, etc.) are then determined as functions of E/N from swarm experiments, where N is the density of the cold, neutral gas. Recovery models based on the Townsend or Raether processes then reduce to determining the gas density at the time of the application of the secondary field. The governing equations are those of gas cooling by conduction, convection, and radiation.

SAHA EQUATION

For high repetition rate switching, the secondary pulse may be applied sufficiently soon after the discharge so that appreciable ionization exists in the gas at high temperature. Equilibrium conditions will be expected to be established after a few collision times. For an arc discharge, the initially high temperature will support a level of ionization which is described by the Saha equation.

Consider the equilibrium state of the general reaction

$$v_1 A_1 + v_2 A_2 \rightleftarrows v_3 A_3 + v_4 A_4 \qquad (10)$$

at constant temperature and pressure. If the partial pressures of the constituents are given by $p_i = x_i p$, then the law of mass action is [12]

$$\frac{\left[x_3^{v_3}\right]\left[x_4^{v_4}\right]}{\left[x_1^{v_1}\right]\left[x_2^{v_2}\right]} p^{\left[v_3+v_4-v_1-v_2\right]} = K(T) , \qquad (11)$$

where $K(T)$ is the equilibrium constant and is given by Nernst's equation

$$\ell n K(T) = \frac{-\Delta H_o}{RT} + \frac{1}{R} \int \left[\int \frac{\left[v_3 C_{p3} + v_4 C_{p4} - v_1 C_{p1} - v_2 C_{p2}\right] dT}{T^2} \right] dT + \frac{\Delta S_o}{R} , \qquad (12)$$

where ΔH_o is the change in enthalpy, ΔS_o is the change in entropy, R is the gas constant, and C_{pi} are the specific heats of the constituents. The x_i's are the mole fractions and are given by

$$x_i = \frac{n_i}{\Sigma n_i} , \qquad (13)$$

where the n_i are the moles of the constituents of the mixture.

For the special case of ionization the reaction is

$$A \rightleftarrows A^+ + e - \epsilon_I , \qquad (14)$$

where ϵ_I is the ionization energy and $v_1 = v_3 = v_4 = 1$, $v_2 = 0$. Define x to be the fractional ionization; that is

$$x = \frac{n_{A^+}}{n_o} , \qquad (15)$$

where $n_o = n_A + n_{A+}$ is the initial particle density. Then $n_A/n_o = (1-x)$ and in equilibrium the total particle density is given by

$$\text{total particle density} = n_A + n_{A+} + n_e = \text{constant} = n_o + n_e. \quad (16)$$

This gives

$$x_1 = \frac{1-x}{1+x}, \quad x_3 = \frac{x}{1+x}, \quad x_4 = \frac{x}{1+x}, \quad (17)$$

and the law of mass action becomes

$$\frac{x^2}{1-x^2} p = K(t), \quad (18)$$

or

$$x = \sqrt{\frac{K(T)}{K(T) + p}}. \quad (19)$$

In the equilibrium constant, ΔH_o gives the energy necessary to ionize 1 mole of atoms. For monatomic gases ($C_{pi} = 5/2R$) the Nernst equation transforms the law of mass action into the Saha equation: namely

$$\frac{x^2}{1-x^2} p = T^{5/2} e^{-\left[\frac{\epsilon_I}{kT} - \frac{\Delta S_o}{R}\right]} = C\, T^{5/2}\, e^{-\frac{\epsilon_I}{kT}}. \quad (20)$$

If the ionized gas is not in steady state equilibrium, then there will be changes in the charged particle densities.

ELECTRON-DENSITY CHANGES

The availability of electrons in the electrode region when the electric field exceeds a certain value (breakdown field) is essential to the breakdown process. That is, the field must exceed a certain value and an electron must be present in this field region at the same time. Irradiated gaps break down at lower average voltages than non-irradiated gaps and rapidly pulse-charged gaps may reach higher voltage levels than slowly charged gaps due to the presence or absence of electrons in the high field region. The presence or absence of electrons in the interelectrode region thus affects the breakdown and recovery. Although the following treatment may seem somewhat abstract from a switch designers viewpoint it is important to the understanding of the physical processes and the time scales involved. Changes in the electron density are usually determined by the continuity equation, which is the zeroth moment of the Boltzmann equation. This was Eq. (2) and is repeated here,

$$\partial_t n_e = \nu_I n_e - \nu_a n_e - \nabla \cdot n_e \mathbf{u} - \alpha_R n_e^2 = \bar{\nu} n_e - \nabla \cdot n_e \mathbf{u} - \alpha_R n_e^2 , \quad (2)$$

where $\bar{\nu} = \nu_I - \nu_a$. Positive (negative) terms on the RHS lead to electron density gains (losses). Without the gradient term the equation becomes the Bernoulli equation

$$\dot{n}_e = \bar{\nu} n_e - \alpha_R n_e^2 , \quad (21)$$

with the solution

$$\frac{1}{n_e} e^{\int \bar{\nu}\, dt} - \int \alpha_R e^{\int \bar{\nu}\, dt}\, dt = C . \quad (22)$$

For the case where ν and α_R are not functions of time and $n(t=0) = n_0$, the solution becomes

$$n_e = \frac{n_0 e^{\bar{\nu} t}}{1 - n_0 \dfrac{\alpha_R}{\bar{\nu}} \left[1 - e^{\bar{\nu} t}\right]} . \quad (23)$$

The special case where $\nu = 0$ gives the solution for recombination

$$n_e = \frac{n_0}{1 + n_0 \alpha_R t} . \quad (24)$$

There are many types of recombination and attachment processes, and the literature covering the coefficients and rates is extensive, see Ref. 13 and its References, for example.

If the gradient term dominates the recombination and attachment terms, the continuity equation becomes

$$\partial_t n_e + \nabla \cdot n_e \mathbf{u} = 0 . \quad (25)$$

The second moment of the Boltzmann equation gives the momentum equation

$$n_e \frac{d}{dt}(m\mathbf{u}) + \nabla \cdot \mathbf{P}^T - n_e \mathbf{F} = -n_e m\nu_m \mathbf{u} , \qquad (26)$$

where $\mathbf{P}^T$ is the pressure tensor, $\mathbf{F}$ is the force, and ν_m is the collision frequency for momentum transfer used in a simple relaxation form for the collision integral. If the pressure is isotropic and thermodynamic equilibrium exists, then the momentum equation can be written as

$$n_e \frac{d}{dt}(m\mathbf{u}) = -\nabla p + n_e \mathbf{F} - n_e m\nu_m \mathbf{u} . \qquad (27)$$

If the electrons are considered inertialess[14], the LHS of the above equation is zero and, for the case in which the force arises from an electric field, the particle flux is given by

$$n_e \mathbf{u} = -\frac{n_e q_e}{m\nu_m} \mathbf{E} - \frac{1}{m\nu_m} \nabla (n_e kT) , \qquad (28)$$

where the ideal gas assumption was used. If the temperature is assumed constant, the flux can be written as

$$n_e \mathbf{u} = -n_e \mu \mathbf{E} - D\nabla n_e , \qquad (29)$$

where $\mu = q_e/m\nu_m$ is the electron mobility and $D = kT/m\nu_m$ is the diffusion coefficient. The continuity equation (25) now becomes

$$\partial_t n_e = \nabla \cdot (n_e \mu \mathbf{E}) + D\nabla^2 n_e \qquad (30)$$

or

$$\partial_t n_e = \mu \mathbf{E} \cdot \nabla n_e + \frac{\rho_c}{\epsilon} \mu n_e + D\nabla^2 n_e , \qquad (31)$$

where ρ_c is the net charge. If there is no E field present, the continuity equation becomes

$$\partial_t n_e = D\nabla^2 n_e . \qquad (32)$$

This is the equation describing diffusion controlled decay (no spatial dependence of temperature). Depending on the geometry of the problem and boundary conditions, many forms of the solution exists.[13,14]

Equation (32) is identical to the equation for thermal diffusion. It is useful to have a solution to this equation for particle or heat diffusion into an infinite medium from an initial particle distribution n(r,t=0) or temperature distribution T(r,t=0), since both processes will occur in recovery. Representing the variables n(r,t) or T(r,t) by Q(r,t), the generic equation becomes, in one dimension,

$$\partial_t Q = D \partial_x^2 Q . \tag{33}$$

Assume that Q(x,t) has an initial distribution given by $Q(x,0) = f(x)$. Introducing a Fourier transform for Q(x,t) as

$$Q(x,t) = \frac{1}{2\pi} \int_{-\infty}^{\infty} F(k,t) e^{ikx} dk$$

$$F(k,t) = \int_{-\infty}^{\infty} Q(x,t) e^{-ikx} dx , \tag{34}$$

the diffusion equation becomes

$$\dot{F}(k,t) = -Dk^2 F(k,t) . \tag{35}$$

Introducing an explicit time dependence of the form

$$F(k,t) = \phi^2(k) e^{-Dk^2 t}$$

it is seen that

$$\phi(k) = F(k,0) = \int Q(x,0) e^{-ikx} dx = \int f(x) e^{-ikx} dx$$

$$F(k,t) = \int f(x) \left[e^{-ikx}\right] \left[e^{-Dk^2 t}\right] dx$$

and the first of Eqs. (34) becomes

$$Q(x,t) = \frac{1}{2\pi} \iint f(x') e^{ik(x-x')} e^{-Dk^2 t} dx' dk .$$

The k integration gives

$$Q(x,t) = \int dx' \, f(x') \frac{1}{\sqrt{4\pi Dt}} e^{-(x-x')^2/4Dt} . \qquad (36)$$

Thus, given the initial spatial distribution $f(x')$ of Q, its value at any position and time can be solved for. It should be noted that the factor

$$\frac{1}{\sqrt{4\pi Dt}} e^{-(x-x')^2/4Dt}$$

is the Greens function $G(x,t;x')$; that is, the solution for a point source in an infinite medium. Substituting $n_e(x,t)$ or $T(x,t)$ for $Q(x,t)$ into Eq. (36) gives the particle density or temperature distribution at any time, t, from their initial distributions $n(x,0)$ or $T(x,0)$. This solution can easily be extended to two and three dimensions.

EFFECT OF SPACE CHARGE

Charges in the electric field (breakdown) region can alter (increase or decrease) the field locally. An understanding of the collective aspects of space charges are essential to the understanding of the breakdown process.

If a small volume of positive charge is introduced between two electrodes, with an overall uniform background field of E_0, then the field in the cathode region will be enhanced and the field in the anode region will be diminished. Denoting the new field strength by $E_0 + E'$, one can from simple breakdown theory* write

$$\frac{\alpha_T}{p} = A e^{-Bp/(E_0 + E')} . \qquad (37)$$

where $B = AV_i$, V_i = ionization potential of gas molecules, and A is a constant which is a property of the gas.

This can be expanded[15] to give

$$\frac{\alpha_T}{p} = \frac{\alpha_0}{p} \left[1 + \frac{BpE'}{E_0^2} + \frac{Bp}{E_0^4} \left[\frac{Bp}{2} - E_0 \right] E'^2 \right] \qquad (38)$$

---

* See Chapter by E. Kunhardt in Vol. II of this series.

where α_o is α_T for $E' = 0$. The change in the number of ion pairs in the whole gap caused by the space charge is then

$$\int_o^d \alpha_T \, dx - \alpha_o d = \alpha_o \frac{Bp}{E_o^2} \int_o^d E' dx + \alpha_o \frac{Bp}{E_o^4} \left[\frac{Bp}{2} - E_o \right] \int_o^d E'^2 dx \,. \tag{39}$$

Since the last integral above is positive, the second term is positive (negative) when E_o/p is less (greater) than $B/2$. The point at $E_o/p = B/2$ is the inflection point in the graph of a/p vs E/p. Thus when $E_o/p < B/2$ (which is the case well to the right of the Paschen minimum), positive space charge in the gap lowers the dielectric strength of the medium.

If the medium is a plasma and a small volume of charge, q_o, exists, then for a Maxwellian distribution the electron and ion densities are

$$n_{e,i} = n_\infty e^{\mp q_e \phi/kT} \,, \tag{40}$$

where n_∞ is the plasma density far away from the volume of charge and ϕ is the potential of the charge distribution. Poisson's equation in spherical coordinates is then[16]

$$\nabla^2 \phi = \frac{1}{r^2} \frac{d}{dr} \left[r^2 \frac{d\phi}{dr} \right] = \frac{n_\infty}{\epsilon_o} q_e \left[e^{q_e \phi/kT_e} - e^{-e_\phi/kT_i} \right] \,. \tag{41}$$

Assume $q_e\phi \ll kT_{e,i}$ then the expansion of the exponentials gives

$$\frac{1}{r^2} \frac{d}{dr} \left[r^2 \frac{d\phi}{dr} \right] = \frac{n_\infty}{\epsilon_o} q_e^2 \left[\frac{1}{kT_e} + \frac{1}{kT_i} \right] \phi \,. \tag{42}$$

The solution is the Debye potential

$$\phi = \frac{q_o}{r} e^{-r/\lambda_D} \,, \tag{43}$$

where λ_D is the Debye length, given by

$$\frac{1}{\lambda_D^2} = \frac{1}{\lambda_e^2} + \frac{1}{\lambda_i^2} \quad \text{where} \quad \lambda_{e,i}^2 = \frac{\epsilon_o kT_{e,i}}{n_\infty q_e^2} \tag{44}$$

Thus the plasma shields the small volume of charge; that is, its vacuum Coulomb potential is diminished by the exponential factor.

These two extreme examples show that charges in an electric field can enhance or diminish that field, locally. Which case occurs depends on the Debye length and number of particles in the Debye sphere. Therefore, a derivation of a breakdown criterion (or determination of the dielectric strength) depends acutely on the "collective" aspects of space charges. If the processes determining a breakdown criterion are strong functions of the ratio of the local electric field to the gas density (E/N), then the gas density N will be determined by the cooling of the gas while the local electric field will be influenced by the character of the space charge. Only when the medium has cooled sufficiently such that the electron mean free path is the dominant parameter in the locally undisturbed field does an avalanche process become valid again and is described by the Paschen Law.

In addition to charged particles affecting the local electric field when the secondary field is applied, they also affect the redistribution of the populations of molecular vibrational states during the post-discharge state. The discharge will set up an initial vibrational state distribution which will be well represented by the Boltzmann Law. Post-discharge conditions will cause, in some cases, the higher levels to become more and more populated at the expense of the lower levels.[17] Upon application of a secondary field during this condition, accelerated electrons will collide with vibrational excited molecules. The influence of these collisions on the Townsend ionization coefficient must be determined for the case of high repetition rate recovery.

In an effort to combine the thermal and electronic character of recovery, yet avoid a microscopic description of discharge re-ignition, the percent recovery can be redefined[18] as the ratio of the instantaneous neutral particle density to the total particle density as $t \to \infty$ in the discharge column; that is,

$$R(t) = \frac{\rho_n(t)}{\rho_c(\infty)}, \qquad (45)$$

where $\rho_n(t)$ is the neutral particle density at time t, and $\rho_c(\infty)$ is the total particle density of the discharge column as $t \to \infty$. Since the total particle density is composed of neutral and ionized species, there results

$$\rho_c(t) = \rho_n(t) + \rho_i(t), \qquad (46)$$

where $\rho_i(t)$ is the density of the charged particle pairs. This leads to a separation of the recovery equation as

$$R(t) = R_H(t) - R_I(t) \tag{47}$$

with

$$R_H(t) = \frac{T_g}{T_c(t)}, \quad R_I(t) = g\frac{\rho_i(t)}{\rho_c(\infty)}, \tag{48}$$

where T_g is the ambient gas temperature, $T_c(t)$ is the temperature of the hot discharge column, and g is a gain factor due to a Townsend avalanche process. $R_H(t)$ represents the fractional recovery due to heat removal while $R_I(t)$ represents the fractional recovery due to charged particle removal. It is interesting that the thermal recovery term, $R_H(t)$, is identical to that derived purely algebraically from the Paschen curve by Rubchinskii.[19] While this treatment is phenomenological, it does allow the instantaneous percent recovery to be written in terms of intrinsic time constants of the recovery processes. However, evaluation of the time constants requires an extensive knowledge of gas and electrode parameters.

EXPERIMENTS

Although gaseous breakdown and discharge developments have been studied since the early nineteen hundreds, it was the mid nineteen hundreds before studies of the general character of the recovery were begun. Rubchinskii[19], in 1969, provided one of the first comprehensive treatments of the recovery of gases after spark discharges at pressures between 150 Torr and about one atmosphere. The major conclusions were:

1. Voltage recovery curves of gases with no metastable atoms or molecules, or mixtures which quickly eliminate the metastables, show two types of behavior:
    a) a short initial section (see schematic diagram of Fig. 3) in which recovery is dominated by an ion layer formed when the secondary voltage is applied.
    b) a long, later section with small slope in which the recovery is dominated by the gas density. This slope is further reduced by the presence of metastable atoms or molecules.

2. Increasing the current pulse length increases the recovery time more than increasing the current amplitude.

3. Pressure (up to about one atmosphere in most cases) has little effect on the recovery time, as does the electrode material.

4. The dielectric recovery of hydrogen is considerably faster than for other gases.

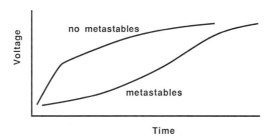

Fig. 3. Generic Recovery Curves

This last conclusion is consistent with the work summarized in the paper of Perkins and Parker,[20] which addresses the results of relevant earlier work.[21-23] These papers report that the dielectric recovery rates depend on the thermal diffusivity of the gas and, for currents of 190 kA and 100 μs duration, temporal variations in α_T, γ and η must be considered to explain the data.

In more recent experimental work[24] recovery after overvolted breakdown of spark gaps at high pressure (100 kPa to 7 MPa) has also been studied. It was found that in the overvolted gap a third section of the recovery curves appeared. This section followed the return to static breakdown (typically 1-100 ms) and could require seconds to return to the original overvolted breakdown level. It was also found that the onset of gas turbulence after the primary discharge depends not only on gas density but on the species as well. For example, hydrogen and helium are frequently non-turbulent at atmospheric pressure (for the first millisecond), but increasing pressure can cause the onset of turbulence. Above several atmospheres air or hydrogen becomes turbulent within 10 μs.

It should be noted that if the switch medium is viewed as a circuit element, then the conductance is important[25]. The conductivity of the medium and the discharge volume are related through the definition

$$G = \int_0^R 2\pi r \sigma dr \, ,$$

where σ is the conductivity and R is the discharge radius.

In a diffuse discharge, stability is maintained between electron production and losses, and heating effects tend to be dominated by gaseous electronic processes. This leads to very fast recovery, especially if the medium contains attaching gases. Generic current wave forms are depicted in Fig. 4. Double-pulse techniques have been used[26] to investigate the recovery of diffuse discharges. While some strongly attaching gases, for example SF_6, recover rapidly, they have trouble maintaining a diffuse discharge. However, some perfluorocarbons, for example C_3F_8, recover rapidly and maintain a diffuse discharge.

CONCLUSIONS

While many studies have been made over the past thirty years on the recovery of dielectric strength of gaseous media, engineering requirements of higher power and greater repetition rates have increased the need for understanding the fundamental processes. Experimental data give broad results and guide lines, but it is uncertain as to how applicable and consistent these results are when applied to different regimes.

In arc discharges most of the recovery must be accomplished through thermal and hydrodynamic processes. In general, these are slow and lead to millisecond recovery times. However, if the properties of the gaseous discharge can be engineered so as to increase the efficiency of the discharge, and thereby reduce the heating, then an improvement in recovery times should result. Likewise, a thorough analysis of the fluid dynamic processes, such as turbulence, may suggest areas of improvement.

In diffuse discharges the recovery must be accomplished by gaseous electronic processes while maintaining the character of the discharge. High field effects on the fundamental discharge processes are important, as well as the partition of energy. Non local-thermodynamic-equilibrium (non LTE) with very high electric fields present as well as energy distributed over electronic,

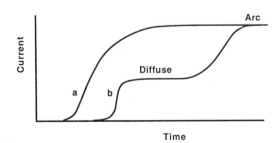

Fig. 4. Current wave forms for non-attaching gases (a) and attaching gases (b).

vibrational, and rotational states requires more detailed Boltzmann or Monte Carlo techniques to describe the recovery of dielectric strength.

REFERENCES

1. J.S. Townsend, The Theory of Ionization of Gases by Collisions, Constabel & Co. Ltd., London (1910).
2. L.B. Loeb, Basic Processes of Gaseous Electronics, Univ. California Press, Berkeley (1961).
3. H. Raether, Electron Avalances and Breakdown in Gases, Butterworths, USA (1964).
4. F. Llewellyn-Jones, Ionization and Breakdown in Gases, Methuen & Co. Ltd, London (1957); The Glow Discharge. Methuen & Co. Ltd., London (1966).
5. J.M. Meek and J.D. Craggs, ed, Electrical Breakdown of Gases, John Willey & Sons, New York (1978).
6. J.D. Cobine, Gaseous Conductors, Dover, New York (1958).
7. L.G. Christophorou, ed, Gaseous Dielectrics, Pergamon Press, New York (IV, 1984, III, 1982, II, 1980).
8. E.E. Kunhardt and L.H. Luessen, ed. Electrical Breakdown and Discharges in Gases, Plenum Press, New York (1983).
9. M. Kristiansen and K. Schoenbach, U.S. Army Workshop on Diffuse Dicharge Opening Switches, (1982) DTIC Report No. AD-A115883;
   M.O. Hagler and M. Kristiansen, DoD Workshop on Repetitive Spark Gap Operation,(1983) DTIC Report No. AD-A132688;
   M. Kristiansen and K. Schoenbach, U.S. Army Workshop on Repetitive Opening Switches, (1981) DTIC Report No. AD-A110770.
10. IEEE Pulsed Power Conferences:
    5th - M.F. Rose and P.J. Turchi, ed, IEEE Publishing Services, New York (1985) IEEE Catalog No. 85C 2121-2;
    4th - T.H. Martin and M.R. Rose, ed, Texas Tech University Press (1983), IEEE Catalog No. 83CH1908-3;
    3rd - A.H. Guenther and T.H. Martin, ed, Texas Tech University Press (1981), IEEE Catalog No. 81CH1662-6;
    2nd - M. Kristiansen and A.H. Guenther, ed, Texas Tech University Press (1979), IEEE Catalog No. 79CH1505-7;
    1st - T.R. Burkes and M. Kristiansen, ed, Texas Tech University Press (1976), IEEE Catalog No. 76CH1147-8.
11. J.A. Rees, Electrical Breakdown in Gases, MacMillan, London (1973).
12. M.W. Zemansky, Heat and Thermodynamics, 5th ed., McGraw-Hill, New York (1957).
13. B.E. Cherrington, Gaseous Electronics and Gas Lasers, Pergamon, NY (1979).
14. S.B. Tanenbaum, Plasma Physics, McGraw-Hill, New York (1967).
15. A. von Engel, Ionized Gases, Oxford Press, Oxford (1965).

16. D.R. Nicholson, <u>Introduction to Plasma Theory</u>, John Wiley and Sons, New York (1983).
17. M. Cacciatore, M. Capitelli, S. DeBenedictis, M. Dilonardo, and C. Gorse, Section 2.3, and M. Capitelli, C. Gorse and A. Ricard, Chapter 11 in <u>Nonequilibrium Vibrational Kinetics</u>, ed. M. Capitelli, Springer-Verlag, NY (1986).
18. R.N. DeWitt, 4th IEEE Pulsed Power Conf., Albuquerque, NM, p. 223, (1983), IEEE Catalog No. 83CH1908-3.
19. A.V. Rubchinskii, in <u>Investigations into Electrical Discharges in Gases</u>, B.N. Klyarfel'd, ed. Pergamon Press, New York (1964).
20. J.F. Perkins and A.B. Parker, <u>J. Appl. Phys.</u>, 41:2895 (1970).
21. R.J. Churchill, A.B. Parker, and J.D. Craggs, <u>J. Electronics and Control</u>, 11:17 (1961).
22. D.E. Poole, A.B. Parker, and R.J. Churchill, <u>J. Electronics and Control</u>, 15:131 (1963).
23. A.B. Parker, D.E. Poole, and J.F. Perkins, <u>Brit. J. Appl. Phys.</u>, 16:851 (1965).
24. S. Moran and S. Hairfield, <u>5th IEEE Pulsed Power Conf.</u>, Alexandria, VA, p. 473 (1985), IEEE Catalog No. 85C 2121-2.
25. J.J. Lowke, R.E. Voshall, and H.C. Ludwig, <u>J. Appl Phys.</u>, 44:3513 (1973).
26. W.W. Byszewski, R.B. Piejak, L.C. Pitchford, and J.M. Proud, GTE Laboratories Inc., Rpt. No. N60921-83-C-(F), 1985.

THE PLASMA EROSION OPENING SWITCH

R.J. Commisso, G. Cooperstein,
R.A. Meger, J.M. Neri, P.F. Ottinger

Plasma Technology Branch
Plasma Physics Division
Naval Research Laboratory
Washington, DC 20375-5000

and B.V. Weber

JAYCOR
Vienna, VA 22180

INTRODUCTION

Inductive energy storage in conjunction with opening switch techniques has many advantages over conventional, capacitive power approaches for high-energy (>10MJ), high-power (>10^{12}W) applications.[1] The principal advantages of inductive storage derive from the 10-100 times higher energy density possible with energy stored in magnetic fields as compared with electric fields and that energy can be stored at low voltage. In theory, this makes compact and economical generators possible.[2] In any inductive storage scheme, the opening switches represent the most critical elements. One such possible switch is the plasma erosion opening switch (PEOS). This switch has been proposed to be used either by itself[3] or in sequence with other opening switches[4] in several inductive-store, pulsed-power systems. Recently, inductive-store/ pulse-compression techniques employing the PEOS have been successfully adapted to conventional pulsed-power generators for the purpose of improving their performance, e.g., peak voltage, peak power, and pulse width.[5-11] In addition to inductive store applications, the PEOS has been used extensively in high power generators for prepulse suppression, improving power flow symmetry, and current risetime sharpening.[5,6,12-18] Depending on the specific application, the PEOS has demonstrated opening times less than 10 ns, conduction times approaching 100 ns, conduction currents as high as 5 MA, and voltage generation over 3 MV without breakdown. Present research is directed toward obtaining a better understanding of PEOS physics and the interaction of the PEOS with the other system components to optimize system performance.

A typical PEOS system is sketched in Fig. 1. Here, the driver may be a conventional pulsed-power generator, a slow capacitor bank (or homopolar generator) with one or more stages of pulse compression prior to the PEOS, or any other appropriate source of current. The PEOS utilizes a source (or sources) of plasma[19,20] located in vacuum between the driver and load. The geometry is such that the plasma is injected between the electrodes so that the plasma is electrically in parallel with the load. The fact that the switch is in vacuum, physically close to the load, helps to minimize the inductance between the load and the switch. Under proper conditions the plasma can conduct the generator pulse to a specified current level, energizing the storage inductance and isolating the load [Fig. 1(a)]. When this current level is reached the switch opens [Fig. 1(b) and (c)] and the generator current is diverted to the load in a time that is short compared with the conduction time.

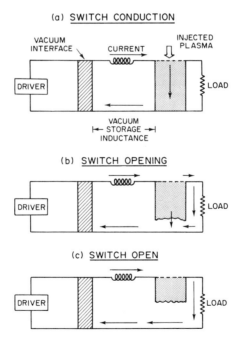

Fig. 1. Schematic representation of Plasma Erosion Opening Switch (PEOS) operation.

The efficiency of current transfer depends strongly on how well the switch plasma parameters are matched to the system parameters (current, switch area, distribution of inductances, etc.).

Physical understanding of the PEOS has evolved over the years as a result of close interplay between theory and experiment. In this article, we first present important aspects of the switch operation from a phenomenological point of view, so that the reader can obtain a sense of the relevant parameter regimes and operational switch characteristics. This discussion is immediately followed by a description of the switch physics. Using this PEOS physics we outline generic PEOS system performance limits. We also review examples of present and future applications of the PEOS. To provide a consistent and convenient data base, we use the results obtained from the Gamble I generator[21] at the Naval Research Laboratory (NRL).

PEOS PHENOMENOLOGY

The general operation of the PEOS can be described with the aid of Fig. 2(a), which is a schematic of the front end of the Gamble I generator configured for inductive-store/pulse-compression experiments with a PEOS.[10] Also shown are a more detailed view of the switch region [Fig. 2(b)] and a typical generator voltage waveform [Fig. 2(c)] measured at the vacuum insulator with an open circuit load. In Fig. 2(a), Gamble I is shown with an electron-beam (e-beam) diode load while in Fig. 2(b) the e-beam diode is replaced with a short circuit. Plasma is injected toward the cathode (inner conductor) through a screen anode. The plasma source consists of three carbon plasma guns[20] equally spaced in azimuth around the inner conductor. The generator is fired a time interval, τ_D, after firing the guns. Negative voltage is applied to the center conductor and current flows through the plasma, energizing the coaxial storage inductance. For the proper combination of plasma parameters the behavior illustrated in Fig. 3 is obtained for a short-circuit load. At some time during the pulse, the generator current, I_G, (measured by Rogowski loop #1) is rapidly diverted (~ 10 ns opening time) to the load, as evidenced by the time history of the load current, I_L, (measured by Rogowski loop #2). An equivalent circuit for this arrangement is shown in Fig 4. Here Gamble I (the driver of Fig. 1) is modeled by a voltage source, V_G [Fig. 2(c) for Gamble I], in series with a generator resistance, R_G (2 ohms for Gamble I). The PEOS is represented by a variable resistance, R_S. The transition inductance, L_T, is the small ($\ll L_0$), parasitic inductance associated with the region between the switch and load and R_L is the load resistance ($R_L = 0$ for a short circuit).

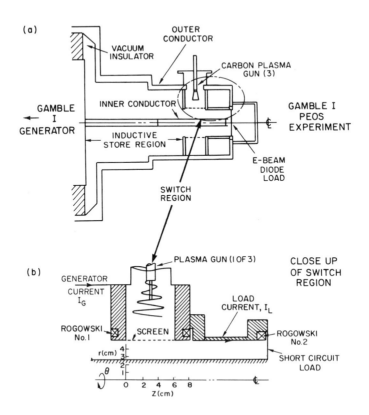

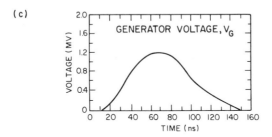

Fig. 2. (a) Schematic of Gamble I PEOS experiment showing relative location of generator, storage inductance, PEOS region, and load. (b) Close-up of switch region showing current diagnostics, electrode polarity, and plasma gun orientation. Here the load is a short circuit. (c) Typical Gamble I generator voltage waveform.

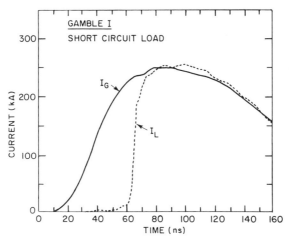

Fig. 3. Measured currents flowing through the generator (I_G) and load (I_L) for Gamble I with a short circuit load.

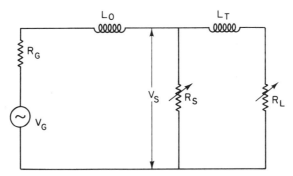

Fig. 4. Generic equivalent circuit for the application of inductive-store/pulse-compression techniques to conventional generators.

For a given generator pulse and load, the switch behavior can change quite markedly by varying the plasma-gun to electrode separation, the time interval, τ_D[22], and the direction of plasma injection with respect to the polarity of the electrodes.[23] All of these factors affect the relevant PEOS plasma parameters at the time the generator is fired.[24,25] With the configuration illustrated in Fig. 2(b), the fastest openings at the highest currents for the Gamble I parameters occur when τ_D is such that the plasma injection speed is > 7 cm/μs, and the switch plasma electron density, n_e, is $\simeq 3 \times 10^{13}$ cm^{-3}. Here the plasma is predominantly C^{++} and the electron temperature is $T_e \simeq 5$ eV. At longer τ_D, n_e

increases to ≳ $5\times10^{14} cm^{-3}$, the plasma no longer has large drift speed toward the cathode, and C^+, neutral carbon, and hydrogen have slowly diffused into the switch region from the walls. Under these conditions the PEOS is observed to conduct longer, but the opening is much slower. For long enough τ_D, the PEOS conducts the entire Gamble I current pulse and never opens. An illustration of this behavior appears in Fig. 5, which is a plot of I_L as a function of time for various values of τ_D obtained from Gamble I with $R_L = 0$ [Fig. 2(b)].

The importance of controlling the direction of plasma injection has been demonstrated by varying the polarity of the electrodes and the injection geometry.[23] Data taken with Gamble I configured as in Fig. 2(b), along with illustrations of the specific plasma injection geometry and electrode polarity, are shown in Fig. 6. The data represent optimized opening switch performance. The PEOS behavior is severely degraded for inward radial injection toward the anode [(Fig. 6(b)] compared with the case of inward radial injection toward the cathode [Fig. 6(a)]. Using a flashboard source, outward radial injection toward the cathode was achieved. This recovers the fast opening behavior of Fig. 6(b), as illustrated in Fig. 6(c).

Much useful information concerning the switch operation can be conveniently obtained using $R_L = 0$; however, most applications of interest require a finite R_L. Some of these applications utilize an e-beam diode for the load. Data[10] from Gamble I using a nominal 12-Ω, e-beam diode load are shown in Fig. 7. The geometry is identical to Fig. 2(a). The load current rises in ≤ 10 ns.

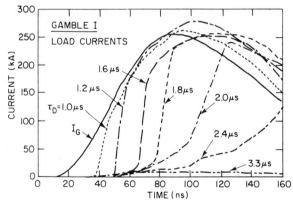

Fig. 5. Plots of load currents, I_L, as a function of time for various time intervals between the plasma gun and the generator firing, τ_D. Also shown is a typical generator current, I_G. Data are from Gamble I with a short circuit load.

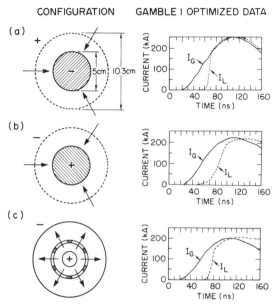

Fig. 6. Illustration of Gamble I system geometry and data for various plasma injection/electrode polarity configurations: (a) inward injection toward cathode, (b) inward injection toward anode and (c) outward injection toward cathode.

However, for this example the peak load current only reaches about 65% of the peak generator current and I_L never reaches I_G, both in contrast to the short circuit case of Fig. 3. This current loss results in a reduction of the maximum power that can be delivered to the load. The cause of the current loss is rooted in the physics of the switch system.[26] As with most opening switches, the PEOS and the system strongly interact in such a way as to adversely affect the switch performance. By understanding the nature of the interaction, however, the losses illustrated in Fig. 7 can be minimized,[26] as shown later.

PRINCIPLES OF OPERATION

The operation of any opening switch can be conveniently divided into two parts: conduction [Fig. 1(a)], and opening [Figs. 1(b) and (c)]. During conduction the switch isolates the load from the rest of the system. In inductive-store applications the switch typically conducts for the time it takes to energize an inductor. When the switch opens, the current is rapidly diverted to the load. Any description of the PEOS operation must explain these switch properties, as well as the rest of the phenomenology described in the previous section.

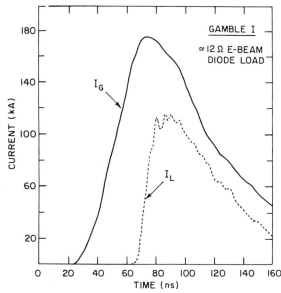

Fig. 7. Measured currents flowing through the generator (I_G) and load (I_L) for Gamble I with an e-beam diode load.

It is important to recognize at the outset that the PEOS conducts a large current (~ MA) across a strong magnetic field (≥ 10 kG) and diverts this current to a finite resistive load using a plasma for which the classical electron-ion momentum transfer time is much longer than either the electron cyclotron period or (in most cases) the time scale of switch operation (~ 100 ns). Also, because of Debye shielding, the voltage across the plasma will appear mostly across a sheath or gap between the plasma and the cathode. The actual flow of current through the plasma is associated with some combination of drift motion and anomalous collisions. One may define an effective "switch resistance" as the ratio of the voltage across the switch to the current flowing through it. However, this quantity will not necessarily be related to the resistivity of the plasma, which determines the fraction of the total energy dissipated by the switch in heating of the switch plasma. Most of the kinetic energy acquired by electrons that traverse the cathode sheath is lost to the anode when those electrons leave the system. The electrical equivalent of the switch behavior will be primarily determined by the cathode sheath physics.

A description of the PEOS operation was introduced, although in a nonswitching context, by Miller et al.,[27] honed by Mendel et al.,[18] significantly modified by Meger et al.,[10] and later described more fully and compared closely with experiment by Ottinger et al.[28] This description is a blend of theoretical concepts and experimental observations. It describes the observed quantitative behavior of the PEOS and has been successfully used to make qualitative predictions. The PEOS physics is briefly summarized in the next paragraph. A more detailed explanation and justifications for assumptions made, along with a discussion of physics constraints on system design, follow later.

The operation is most easily described as a sequence of four phases: conduction, erosion, enhanced erosion, and magnetic insulation. The last three phases constitute the switch opening. Effects related to $\overline{JXB}$ body forces and high density plasma associated with plasma-wall interaction or field emission are neglected here but will be discussed in what follows. The plasma is injected into the switch region and voltage applied to the cathode. As long as the switch current density remains below a predictable value, which depends on only the plasma parameters, the plasma acts as a good conductor [Fig. 1(a)]. The conduction occurs through a non-neutral region, called a sheath or gap, at the cathode in a bipolar space-charge-limited fashion with the electron component emanating from the cathode and the ion component provided by the injected plasma. When the switch current density becomes high enough that the bipolar space-charge condition cannot be satisfied by the ions from the injected plasma, the gap widens, providing more ions. This is called the erosion phase. When the switch current increases to the point where the average electron Larmor radius is comparable with the gap size, the electron lifetime in the gap increases and as a result the space-charge condition is modified in such a way that even more ions are required. This is called the enhanced erosion phase and it is during this phase that the gap widens very quickly. A high voltage is generated across the gap and a substantial fraction of current is diverted to the load. The switch is totally open when the magnetic insulation phase is reached. This occurs at a value of current for which the average Larmor radius is less than the gap size and all the current reaches the load [Fig. 1(c)]. A more detailed explanation now follows.

SWITCH PHYSICS

Switch Conduction. As described in the previous section, a plasma is injected into the switch region and after a time, τ_D, the generator is fired, applying a high voltage to the cathode, $- V_C$. This voltage causes a non-neutral region or gap of dimension D_o (typically, $D_o \gg$ Debye length characteristic of the injected plasma) to be formed at the cathode surface. The situation at one instant during the switch conduction is schematically illustrated in Fig. 8. Initially, the only current flowing through the gap will be ion current associated with the flux of ions from the injected plasma. Very soon, however, the electrostatic field at the cathode becomes large enough[29] that field emission[30] of elec-

and experimental observations. It describes the observed quantitative behavior of the PEOS and has been successfully used to make qualitative predictions. The PEOS physics is briefly summarized in the next paragraph. A more detailed explanation and justifications for assumptions made, along with a discussion of physics constraints on system design, follow later.

The operation is most easily described as a sequence of four phases: conduction, erosion, enhanced erosion, and magnetic insulation. The last three phases constitute the switch opening. Effects related to JXB body forces and high density plasma associated with plasma-wall interaction or field emission are neglected here but will be discussed in what follows. The plasma is injected into the switch region and voltage applied to the cathode. As long as the switch current density remains below a predictable value, which depends on only the plasma parameters, the plasma acts as a good conductor [Fig. 1(a)]. The conduction occurs through a non-neutral region, called a sheath or gap, at the cathode in a bipolar space-charge-limited fashion with the electron component emanating from the cathode and the ion component provided by the injected plasma. When the switch current density becomes high enough that the bipolar space-charge condition cannot be satisfied by the ions from the injected plasma, the gap widens, providing more ions. This is called the erosion phase. When the switch current increases to the point where the average electron Larmor radius is comparable with the gap size, the electron lifetime in the gap increases and as a result the space-charge condition is modified in such a way that even more ions are required. This is called the enhanced erosion phase and it is during this phase that the gap widens very quickly. A high voltage is generated across the gap and a substantial fraction of current is diverted to the load. The switch is totally open when the magnetic insulation phase is reached. This occurs at a value of current for which the average Larmor radius is less than the gap size and all the current reaches the load [Fig. 1(c)]. A more detailed explanation now follows.

SWITCH PHYSICS

Switch Conduction. As described in the previous section, a plasma is injected into the switch region and after a time, τ_D, the generator is fired, applying a high voltage to the cathode, V_C. This voltage causes a non-neutral region or gap of dimension D_0 (typically, $D_0 \gg$ Debye length characteristic of the injected plasma) to be formed at the cathode surface. The situation at one instant during the switch conduction is schematically illustrated in Fig. 8. Initially, the only current flowing through the gap will be ion current associated with the flux of ions from the injected plasma. Very soon, however, the electrostatic field at the cathode becomes large enough[29] that field emission[29] of elec-

THE PLASMA EROSION OPENING SWITCH

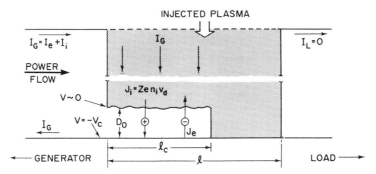

Fig. 8. Schematic of switch operation during the conduction phase.

trons occurs at the cathode. Electron and ion currents are then conducted across the gap in a bipolar space-charge-limited fashion[31]. The ion current density at the plasma-gap boundary is

$$J_i = Zen_i v_d, \qquad (1)$$

where n_i is the ion density in the plasma, assumed uniform and constant during conduction and opening, e is the electronic charge, and Z is the ion charge state. The ion drift speed, v_d, is the average component of ion velocity toward the cathode surface for ions entering the gap. It is usually approximated by the average plasma injection speed. This may be an over estimate because geometry effects and possible plasma wall interactions can decrease the actual ion collection rate. Thus, at any time during conduction the ion current associated with the injected plasma is simply the integral of J_i over the area through which current is flowing. In cylindrical geometry we have

$$I_i \simeq 2\pi r_c \ell_c(t) J_i . \qquad (2)$$

Note that the switch area is a function of time. For bipolar space-charge-limited flow,[31]

$$\frac{I_i}{I_e} = (Zm_e/m_i)^{1/2}, \qquad (3)$$

where I_e is the electron current, $m_e(m_i)$ is the electron (ion) mass, and relativistic effects are neglected. Thus, for C^{++}, $I_e/I_i \simeq 100$, and the total current flowing in the switch $I_S = I_e + I_i \simeq I_e$.

The conduction (and opening) phase has been studied in detail by using magnetic probes[32] to obtain spacially and temporally resolved measurements of the magnetic field inside the PEOS in the cylindrical geometry of Fig. (2). These measurements suggest that as the switch current increases, the current channel width at the cathode, ℓ_C in Fig. 8, also increases. This increase is predicted by the space charge condition of Eq. (3) with the ion current given by Eq. (2). In the cylindrical geometry, we have,[32] using Eq. (2) in Eq. (3),

$$\ell_c(t) \simeq \frac{I_S(t)}{(m_i/Zm_e)^{1/2} 2\pi r_c J_i} . \qquad (4)$$

Because J_i is constant, Eq. (4) predicts that the switch current density ($\simeq I_S/2\pi r_c \ell_c(t)$) will be also constant. This effect is observed.[32] As the switch current increases, the current channel in the plasma is observed to broaden as rapidly as $\ell_c(t)$. The current channel is much larger than the collisionless skin depth, c/ω_{pe}, where c is the speed of light and ω_{pe} is the electron plasma frequency. This implies a resistivity for the plasma during the conduction phase that is much larger than classical and results in a predominantly radial current flow.

The switch will continue to conduct in the manner depicted in Fig. 8 as long as Eq. (4) can be satisfied, i.e., for $\ell_c < \ell$. However, because the generator forces I_S to increase in time while the ion current density is fixed, the switch current eventually reaches a value where the ion current density integrated over the full switch area becomes insufficient to satisfy Eq. (4), i.e., $\ell_c = \ell$. This occurs when

$$I_S(t) = (m_i/Zm_e)^{1/2} Zen_i v_d A , \qquad (5)$$

where A is the maximum current channel area ($\simeq 2\pi r_c$ for cylindrical geometry) allowed by the switch geometry. At this point the conduction phase ends. A voltage consistent with the Child-Langmuir law, $V_S^{3/2} \propto D_0^2 I_S$, appears across the switch and, (depending on L_T and R_L), on the load. For the plasma parameters quoted earlier, $J_i \simeq 30$ A/cm^2. If A ~ 1000 cm^2, Eq. (5) predicts that a current as high as I_S ~ 3 MA may be conducted in this fashion.

Note that in this description of the conduction phase (Fig. 9) there is no $\overline{J}\times\overline{B}$ force in the direction of power flow on the load side of the switch plasma until the current channel reaches it. If the opening occurs very rapidly after this instant, the current no longer flows between the electrodes and $\overline{J}\times\overline{B} \simeq 0$ in the direction of power flow. Some $\overline{J}\times\overline{B}$ related motion on the driver side of the switch plasma will occur as soon as current flows through the switch. This can result in a compression of the plasma and thus an increase in the plasma density, which may affect the switch opening, as discussed next in the section on switch opening. If for this or some other reason the opening occurs very slowly after the current channel broadens to the load side of the switch, the switch plasma will be accelerated in the direction of power flow and a situation similar to that of a plasma accelerator will exist. For this situation, a different regime of physics (which will not be discussed here) is relevant. The switch should be designed so that the fast opening case is obtained.

In this picture of conduction, the (small) ion current associated with the injected plasma controls the (large) current the switch will conduct before opening. Thus, for a given A and Z the conduction is essentially controlled by the directed ion flux density to the cathode, $\Gamma_i \equiv n_i v_d$. If other processes that provide additional ion flux become dominant, e.g., as a result of plasma-wall interaction or arc phenomena, control of not only conduction but also the rapid opening may be lost. Fortunately, these processes usually take a long time (compared with the time-scale for switch operation) to manifest themselves. Control of Γ_i is achieved for the carbon plasma guns discussed previously by varying τ_D, as suggested earlier (Fig. 5). In Fig. 9 a more quantitative demonstration of this control is displayed.[22] Using Gamble I configured as in Fig. 2(b), the maximum switch current I_S ($I_S \equiv I_G - I_L$) was measured as a function of τ_D. The ion density and drift velocity at different τ_D were estimated from double floating probes and Faraday cups and used in a circuit interactive computer code that models the entire switch opera-

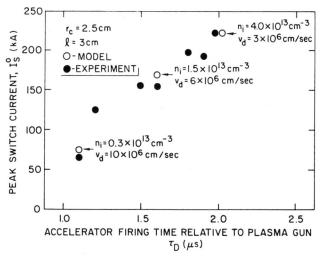

Fig. 9. Plot peak switch current, $I_S(I_S \equiv I_G - I_L)$, as a function of relative delay, τ_d. Also shown are numerical results from the model, using the measured values of n_i and v_d.

tion.[28] The final current channel width, ℓ, was estimated from the magnetic field measurements.[32] The model, which incorporates the current limit of Eq.(5), and experimental results agree.

Switch Opening. The switch opening can be described in three phases: erosion, enhanced erosion, and magnetic insulation. These three phases are schematically depicted in Fig. 10. When the current limit of Eq.(5) is reached the plasma must respond in such a way that the space charge condition of Eq.(3) remains satisfied. This is accomplished by the gap size increasing.[18,27,33,34] That is, the boundary between the plasma and the non-neutral gap recedes from the cathode, exposing new ions. This process is termed "erosion" and it increases the rate at which ions enter the gap. In the frame of the plasma boundary, the ions acquire an additional drift speed equal to dD/dt, where D(t) is the gap size determined by

$$I_i = Zen_i A(v_d + \frac{dD}{dt}) \qquad (6)$$

and I_i is given by Eq. (3). As the gap increases in response to the increasing electron current the voltage across the gap (switch) also increases (because of the Child-Langmuir law) as $V_S^{3/2} \propto D^2 I_S$. If this voltage drives current through the load, we may say that the switch begins to open [Fig. 10(a)].

(a) EROSION PHASE

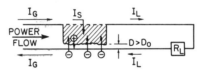

(b) ENHANCED EROSION PHASE

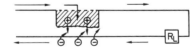

(c) MAGNETIC INSULATION PHASE

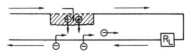

Fig. 10. Schematic of switch opening: (a) erosion phase, (b) enhanced erosion phase, and (c) magnetic insulation phase.

THE PLASMA EROSION OPENING SWITCH

Until this point the effects of the magnetic field associated with I_S have been ignored. These effects become important when the typical electron Larmor radius becomes comparable with the gap size. For cylindrical geometry, particle-in-cell (PIC) code simulations show that magnetic field effects become important when I_S reaches a critical value given by[35]

$$I_S \simeq I_C \equiv 1.36 \times 10^4 (\gamma^2 - 1)^{1/2} \frac{r_c}{D} \qquad (7)$$

where I_C is in amperes and γ is the ratio of the electron energy to its rest energy. (The PIC code results differ only slightly from simple analytic estimates.)

When I_S approaches I_C of Eq. (7), the electron flow across the gap changes and the electron space-charge distribution in the gap is altered. Electrons now $\bar{E} \times \bar{B}$ drift along the switch length in the self-consistent electric and magnetic fields in the gap, traveling a distance characterized by the length of the switch, $\sim \ell$, rather than just crossing the gap, D, before reaching the plasma. An important consequence associated with this alteration of the electron space charge is a modification of the bipolar space-charge-limited flow condition. The electron space charge near the plasma surface is increased and the required ion current is greatly enhanced[36] over that given in Eq. (3). The new condition is[36]

$$\frac{I_i}{I_e} \approx [2m_e Z(\gamma + 1)/m_i]^{1/2} \left(\frac{\ell}{D}\right) \qquad (8)$$

and the gap must open faster to provide the additional required ion current. This is called the enhanced erosion phase [Fig. 10(b)] and is responsible for the observed rapid switch opening. Equation (8) now determines I_i in Eq. (6) and erosion rates of $dD/dt \sim 10^8$ cm/s can be obtained. For $\gamma^2 \gg 1$, Eq. (7) gives $V_S \propto I_S D$. Thus, as D increases, V_S also increases, forcing more current through the load.

Depending on the geometry, the magnetic field may vary along the switch length. Thus, electrons flowing across one part of the switch gap may experience a smaller magnetic field than those flowing across another part of the gap. When the load current becomes large enough that the magnetic field prevents electrons from traversing the gap anywhere, the switch is said to be magnetically insulated. This is the last phase of the switch opening and is depicted in Fig. 10(c). When $I_L \geq I_C$, the electrons are completely insulated so that $I_e = 0$, $I_S = I_i \ll I_C$, and the switch is open. The ion current is no longer enhanced because the magnetic field now holds the electron trajectories close to the cathode surface and I_i is simply the one species space-charge-limited current for ions.

To summarize the PEOS model, the PEOS remains closed during the conduction phase with the switch area and ion flux density from the injected plasma determining the peak switch current. The switch opening commences with the erosion phase, accelerates during the enhanced erosion phase and ends with the magnetic insulation phase. The transition of I_i from bipolar flow [Eq. (3)] to enhanced flow [Eq. (8)] back to single species flow can be modeled in a continuous way.[37] A comparison of the measured time histories of I_G and I_L for Gamble I (nominal 10-Ω, e-beam diode load) with those computed using this model is shown in Fig. 11. The model correctly predicts the observed conduction and opening characteristics.

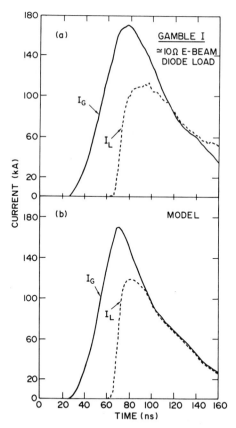

Fig. 11. Comparison of measured and calculated time histories for generator (I_G) and load (I_L) currents.

Throughout this discussion $\bar{J}X\bar{B}$ forces have been assumed to be negligible in opening the gap. In the coaxial geometry of Fig. (2), magnetic field measurements[32] reveal that the current predominantly flows in the radial direction, suggesting small radial $\bar{J}X\bar{B}$ forces. Moreover, calculations for coaxial geometry that assume the current flows in a thin channel parallel to the axis before going radially to the cathode result in a time history and density scaling for opening that is not supported by experiment. The description of the conduction phase presented earlier asserts that $\bar{J}X\bar{B}$ forces perpendicular to the current flow will have negligible effect on the switch behavior as long as the current channel length increases more rapidly and the switch opens faster than the time for these forces to significantly accelerate the plasma. The magnetic field measurements on Gamble I indicate that axial $\bar{J}X\bar{B}$ forces are not significant,[32] in accordance with this assertion. However, experiments are needed to more fully explore the long conduction time (> 100 ns), high current ($\gtrsim$ MA) regime.

As discussed earlier in this section, for a given A and Z the directed ion flux density to the cathode, $\Gamma_i = n_i v_d$, controls how much current the switch will conduct before the erosion phase begins. How well the switch opens, on the other hand, will be determined by the gap size relative to the characteristic electron Larmor radius. Assuming $v_d \ll dD/dt$, Eq. (6) gives

$$\frac{dD}{dt} \cong \frac{I_i}{Zen_i A} . \qquad (9)$$

Thus, for a given value of I_i, A, and Z, the rate at which the gap opens is $\propto n_i^{-1}$. To achieve a high conduction current $\Gamma_i A$ must be large [Eq.(5)] - but to achieve rapid opening and large final gap size n_i must be small [Eq.(9)]. This implies v_d must be large to have both high conduction current and good opening characteristics. Present plasma sources[19,20] have $Ze\Gamma_i \simeq 30$ A/cm^2 with $n_i \simeq 10^{13}$ cm^{-3} and $v_d \simeq 10^7$ cm/s. An important area of research is the development of plasma sources with similar or greater Γ_i but with lower n_i. For a given switch configuration, the optimum lower bound on n_i is determined from $\bar{J}x\bar{B}$ force considerations. The effect of varying n_i and v_d while keeping Γ_i fixed is illustrated in Fig. 12, which is a result of a computation for Gamble I parameters using the model just described.[28,38] Although the current at which the switch begins to open remains the same, the best switch performance is obtained with the highest v_d and lowest n_i [Fig. 12(a)]. This behavior has been confirmed in the laboratory. The plasma injection geometry will also affect the value of Γ_i, which explains the data in Fig. 6. For example, in Fig. 6(b), the flow velocity is toward the anode, resulting in v_d and n_i at the cathode that are different from that illustrated in Figs. 6(a) and (b). Control of the PEOS behavior is strongly dependent on the "quality"

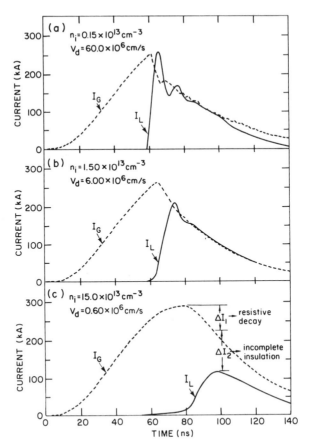

Fig. 12. Results of numerical computations based on the PEOS model showing the effects of varying the values of n_i and v_d while keeping the product fixed. The load is a fixed 6Ω resistor.

of the directed flux to the cathode - a "high quality flux" having, for a given $\Gamma_i A$, a low n_i and high v_d.

SYSTEM CONSIDERATIONS

In order for the switch to operate in the most efficient way (i.e., for as much of the stored inductor current to reach the load as is permitted by magnetic flux conservation) two conditions must be met. First, the switch must open in a time that is fast compared with the characteristic resistive decay time of the current during opening. This is equivalent to saying that magnetic flux be very nearly conserved during the opening. Second, the electrons in the gap must be totally insulated. The final switch gap must be large compared with the characteristic electron Larmor radius in the gap or, equivalently, $I_L \gtrsim I_C$. If these two conditions are not met, then the situation depicted in Fig. 12(c) may arise. There, the generator current at the time of peak load current is much less than the peak generator current [ΔI_1 of Fig. 12(c)] because the opening is slow compared with the current decay time during opening. A reduction in current of only 5% would be expected from magnetic flux conservation. There is also a loss of current to the load as evidenced by the finite difference between generator and load current after opening [ΔI_2 of Fig. 12(c)], resulting from incomplete insulation. This loss is also evident in Fig. 7, where $I_G > I_L$ after opening, indicating that current is still flowing through the switch.

These two conditions can be combined to define an operational window for PEOS systems. An example of a result of this analysis[39] for the Gamble I generator is shown in Fig. 13. Here the accessible load voltages, V_L, are plotted as a function of the load resistance, R_L, for an assumed switching time, $\tau_S \sim I_L(dI_L/dt)^{-1}$, storage inductance, L_o, and ratio of cathode radius to final gap size, r_c/D. The dotted line in Fig. 13 represents the maximum voltage that can be generated at the load, $V_L = R_L I_L$, consistent with resistive decay of the current during opening. The characteristic resistive decay time for the current is $\sim L_o/R$, where R is some effective system resistance during opening. The voltage, V_L, can not be higher than this dotted line for a given R_L. Here, how efficiently the generator energizes the storage inductance has also been included. This efficiency can be as high as 80%, depending on the value of L_o. The solid line is obtained by requiring magnetic insulation, i.e., $I_L(=V_L/R_L) \approx I_C$. To minimize current loss, the switch electrons must be insulated in the gap. For a given R_L, I_L must be large enough that V_L lie at or above the solid line. Thus for a given τ_S, L_o, and r_c/D the region between the dotted and solid lines (shaded portion of Fig. 13) represents an operational window for efficient system operation.

If D is increased (for a given r_c), it takes less current to insulate the electrons and therefore V_L will be smaller for a given R_L. This shifts the solid line in Fig. 13 to the right and down from the one shown. If τ_S is decreased (for a given L_o) the maximum current available to the load for a given R_L increases because the switching time becomes small compared with L_o/R. This increases the maximum accessible V_L for a given R_L, resulting in the dashed line in Fig. (13) shifting up from the one shown. Thus the system is optimized by making D as large and τ_S as small as possible. In practice, D and τ_S are themselves optimized through proper choice of plasma parameters (small n_i, large v_d) as discussed earlier [Eq. (9)].

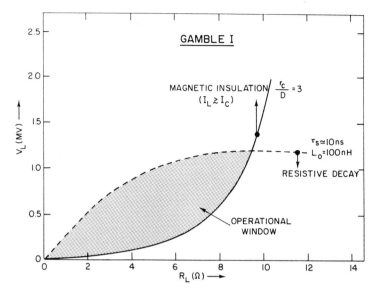

Fig. 13. Operational window for a PEOS system on the Gamble I device for $L_0 = 100$ nH, $r_c/D = 3$, and $\tau_S = 10$ ns. Regime of efficient operation (minimum current loss) is area between dotted and solid line.

This analysis illustrates the complex interactive nature of this switch system. In general, the strong interaction between the pulsed power system and opening switch is a characteristic of most inductive store systems, regardless of the particular opening switch employed. By properly designing the system and choosing the switch plasma parameters, current loss for a given application can be eliminated or at least minimized. Other loss mechanisms not addressed in the foregoing analysis are possible. For example, current loss can occur in the transition region between the switch and load. This effect will not be discussed here except to say that it has been shown experimentally that this loss can also be minimized.[26]

APPLICATIONS

The PEOS has a wide variety of applications in fields of research and development that require power conditioning at high power levels (≥ 1 TW). Three applications of current interest will be briefly reviewed here.

One of the most exciting applications involves the adaptation of a PEOS inductive-store/pulse-compression system to state-of-the-art generators to produce a higher power, higher voltage, shorter risetime pulse than can otherwise be obtained from the generator.

THE PLASMA EROSION OPENING SWITCH

This application is loosely termed power and voltage multiplication and has particular relevance in Light Ion Beam Inertial Confinement Fusion (LIBICF) research.[5-9] Fast risetime and high voltage are particularly attractive for LIBICF because of the required short pulses (10-20 ns) and ion energies as high as $\simeq$ 30 MeV (for lithium ions).

The potential for significant power and voltage multiplication was first demonstrated by Meger et al.[10] The data in Fig. 14 serve to illustrate the salient features of this application, albeit under non-optimized conditions. The data shown are the power to the load, P_L, the load voltage, V_L, and the energy delivered to the load, E_L, as functions of time for three cases: (1) Gamble I configured with the PEOS inductive-store/pulse-compression system shown in Fig. 2(a) and an e-beam diode load of nominally 6Ω; (2) Gamble I configured with a matched, 2Ω, constant resistance load (as a practical matter, a minimum of $\simeq$ 30 nH separates the generator from the load in this configuration); and (3) Gamble I configured with 6Ω, constant resistance overmatched load (again with $\simeq$ 30 nH of inductance between the generator and load). The data for the first case were obtained from a Gamble I shot. The measured generator voltage waveform for that particular shot was used to calculate the parameters for the other two cases. The risetimes of both P_L and V_L with the PEOS system are $\simeq$ 10 ns, compared with $\simeq$ 50 ns for the conventionally figured systems. The voltage, V_L, is a factor ~ 2 higher with the PEOS system compared with the matched load and slightly higher than the overmatched case; however, P_L is twice as high with the PEOS system compared with the overmatched case. For case 1, a current loss (discussed earlier) of $\simeq$ 23% was observed, which accounts for the peak P_L with the PEOS being comparable with the matched load case. Power multiplication over the peak matched load power of ~ 1.5 could have been obtained at ~ 1.4 times the overmatched load voltage if all the current went to the load.

The cost associated with achieving power and voltage multiplication is illustrated in the bottom graph of Fig. 14. For the (non-optimized) shot illustrated, the PEOS system delivers $\simeq$ 66% of the energy that could be delivered to a matched load. Note, however, that in the first 20 ns of the pulse, the PEOS system delivers twice as much energy to the load as the matched load case during the same time interval. There is presently a research effort at NRL directed toward improving the current transfer so that significant power and voltage multiplication can be obtained with a minimum cost in energy. This effort focuses on: (1) the development of new plasma sources which can provide a higher quality flux than is presently available from the sources now used and (2) making best use of the present sources through optimization of the system geometry. Recent calculations suggest that the PEOS system could give a factor ~ 2 power multiplication and ~ 4 voltage

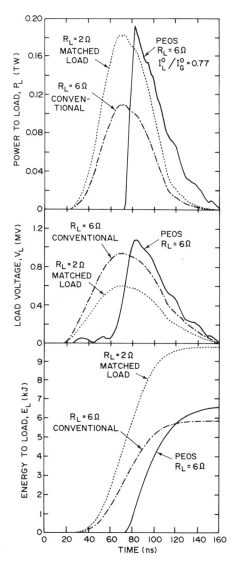

Fig. 14. Sample data from Gamble I illustrating some advantages that a generator configured with a PEOS inductive store/pulse compression system has over a conventionally configured generator with both a matched load (2Ω) and an overmatched load (6Ω).

multiplication over matched load values with the present sources for Gamble I parameters if the proper system geometry could be used.

With the hope of attaining some degree of power multiplication, initial PEOS experiments on the Particle Beam Fusion Accelerator I (PBFA I) at Sandia National Laboratories were performed.[5,6,9] In these experiments the PEOS was additionally used to minimize the effects of intermodule jitter, symmetrize the power flow, and narrow the pulse width. The experiments were successful in the latter three areas. Current loss in the switch and between the switch and load prevented increased power to the load.[5,6] The PBFA II system,[6,9] which is now being tested, was designed from the start to take advantage of inductive-store/pulse-compression techniques. For PBFA II, the PEOS must transfer $\simeq$ 6 MA in $\simeq$ 10 ns to generate $\simeq$ 30 MV in a 5-Ω load. Calculations indicate that with proper system design and a small extension of present plasma source technology these requirements can be met.

Another application involves adapting the PEOS to a conventional generator, primarily to reduce prepulse, provide a shorter rise-time pulse, and improve the symmetry of the power flow. Here, a large vacuum inductive store region is not necessary. In addition to the impact on ICF discussed in the previous paragraph, this application has particular relevance to ion diode physics[13,15,16] and the generation of imploding z-pinch plasmas.[12,14,17] Experiments at Physics International Corporation[14,17] using imploding wire array and at NRL[12] using annular gas puffs indicate an improvement in the quality of the implosions (better axial uniformity, smaller pinch radius, etc.) when the PEOS is used. The NRL group also reports that K-line radiation from neon pinches can be up to 2.5 times higher with the PEOS than without the PEOS at the same current level.[12] Although a detailed understanding of these phenomena is still forthcoming, the preliminary experimental results are encouraging.

The final application considered here, and potentially the one with the most impact on the development of pulsed power, focuses on the eventual elimination of conventional generators from the pulsed-power system altogether. Such generators usually employ a Marx type configuration of capacitors as a prime energy store in conjunction with a water (or oil) dielectric pulse-forming line. They require large quantities of oil and high-purity water for high voltage insulation and they employ a high voltage(> 1 MV) vacuum interface. These generators tend to be very large and do not scale economically to high energy ($\gtrsim$ 20 MJ) systems. One alternative system utilizes either a low voltage($\sim$ 50 kV), compact, high energy density capacitor bank [4,40,41] or a homopolar generator[2] to slowly ($10-10^3$ μs) energize a vacuum inductor, after which a series of opening switches successively increases the voltage and decreases the pulse risetime. A critical element in this program is the development of the required opening switch technology. The PEOS is a prime candidate for the final opening switch in such a system.

This work is just coming out of its infancy and should provide interesting results in the near future.

The authors would like to thank F.C. Young, I.M. Vitkovitsky, J.D. Sethian and R.F. Fernsler for their helpful comments. This work was supported by Sandia National Laboratories, the Department of Energy, the Defense Nuclear Agency and the Office of Naval Research.

REFERENCES

1. S.A. Nasar and H.H. Woodson, Storage and Transfer of Energy for Pulsed-Power Applications, Proceedings of the Sixth Symposium on Engineering Problems of Fusion Research, San Diego, CA, IEEE Pub. No. 75H1097-5-NPS, (1975).
2. R.D. Ford, D. Jenkins, W.H. Lupton, and I.M. Vitkovitsky, Pulsed High-Voltage and High-Current Outputs from Homopolar Energy Storage System, Rev. Sci. Instrum., 52:694 (1981).
3. K.D. Bergeron and J. Pace VanDevender, Fast Opening Plasma Switch, 1978 IEEE International Conference on Plasma Science, Monterey, CA, (1978), IEEE Cat. No. 78CH1357-3NPS.
4. P.F. Ottinger, R.J. Commisso, W.H. Lupton, and J.D. Shipman, Jr., System Study for an Inductive Generator, Proc. Fifth IEEE Pulsed Power Conference, Washington, DC (1985), IEEE Cat. No. 85C2121-2.
5. R.W. Stinnett, W.B. Moore, R.A. Meger, J.M. Neri, and P.F. Ottinger, Plasma Opening Switch Experiment on PBFA-I, Bul. Am. Phys. Soc., 29:1207 (1984).
6. J.P. Vandevender et al. (48 authors), Light-Ion Fusion Research in the U.S.A., Tenth International Confrence on Plasma Physics and Controlled Nuclear Fusion Research, London, U.K. (1984), IAEA, Vienna, 3:59 (1985).
7. K. Imasaki et al. (20 authors), Light-Ion Fusion Research in Japan, Tenth International Conference on Plasma Physics and Controlled Nuclear Fusion Rearch, London. U.K. (1984), IAEA, Vienna, 3:71 (1985).
8. R.A. Meger, J.R. Boller, R.J. Commisso, G. Cooperstein, S.A. Goldstein, R. Kulsrud, J.M. Neri, W.F. Oliphant, P.F. Ottinger, T.J. Renk, J.D. Shipman, Jr., S.J. Stephanakis, B.V. Weber, and F.C. Young, The NRL Plasma Erosion Opening Switch Research Program, Fifth International Conference on High-Power Particle Beams, San Francisco, CA (1983).
9. J. Pace VanDevender, Light Ion Beam Fusion, Fifth International Conference on High Power Particle Beams, San Francisco, CA (1983).

10. R.A. Meger, R.J. Commisso, G. Cooperstein, and Shyke A. Goldstein, Vacuum Inductive Store/Pulse Compression Experiments on a High Power Accelerator Using Plasma Opening Switches, Appl. Phys. Lett., 42:943 (1983).
11. R.A. Meger, J.R. Boller, D. Colombant, R.J. Commisso, G. Cooperstein, S.A. Goldstein, R. Kulsrud, J.M. Neri, W.F. Oliphant, P.F. Ottinger, T.J. Renk, J.D. Shipman, Jr., S.J. Stephanakis, F.C. Young, and B.V. Weber, Application of Plasma Erosion Opening Switches to High Power Accelerators for Pulse Compression and Power Multiplacation, Fourth IEEE Pulsed Power Conference, Albuquerque, NM, (1983), (IEEE Cat. No. 83CH1908-3).
12. S.J. Stephanakis, S.W. McDonald, R.A. Meger, P.F. Ottinger, F.C. Young, C.G. Mehlman, J.P. Apruzese, and R.E. Terry, Effect of Pulse-Sharpening on the Gas-Puff Imploding Plasma, Bull. Am. Phys. Soc., 29:1232 (1984).
13. S.J. Stephanakis, T.J. Renk, G. Cooperstein, Shyke A. Goldstein, R.A. Meger, J.M. Neri, P.F. Ottinger, and F.C. Young, Effect of Pulse Compresion on Ion Beams from Pinch Reflex Diodes, Bull. Am. Phys. Soc., 28:1055 (1983).
14. R. Stringfield, P. Sincerny, S.-L. Wong, G. James, T. Peters, and C. Gilman, Continuing Studies of Plasma Erosion Switches for Power Conditioning on Multiterawatt Pulsed Power Accelerators, IEEE Trans. Plasma Sci., PS-11:200 (1983).
15. R.A. Meger and F.C. Young, Pinched-Beam Ion-Diode Scaling on the Aurora Pulser, J. Appl. Phys., 53:8543 (1982).
16. W.F. Oliphant, S.J. Stephanakis, G. Cooperstein, Shyke A. Goldstein, R.A. Meger, D.D. Hinshelwood, and H.U. Karow, Small Diameter Pinch-Reflex Diode Behavior with Plasma Erosion Switch Beam Front Sharpening, 1981 IEEE International Conference on Plasma Science, Santa Fe, NM, (1981), IEEE Cat. No. 81CH1640-2NPS.
17. R. Stringfield, R. Schneider, R.D. Genuario, I. Roth, K. Childers, C. Stallings, and D. Dakin, Plasma Erosion Switches with Imploding Plasma Loads on a Multiterawatt Pulsed Power Generator, J. Appl. Phys., 52:1278 (1981).
18. C.W. Mendel, Jr. and S.A. Goldstein, A Fast-Opening Switch for Use in REB Diode Experiments, J. Appl. Phys., 48:1004 (1977).
19. T.J. Renk, S.J. Stephanakis, R.A. Meger, and J.R. Boller, Plasma Erosion Opening Switch Experiments Using Flashboard Sources, Bull. Am. Phys. Soc., 28:1055 (1983).
20. C.W. Mendel, Jr., D.M. Zagar, G.S. Mills, S. Humphries, Jr., and S.A. Goldstein, Carbon Plasma Gun, Rev. Sci. Instrum., 51:1641 (1980).
21. G. Cooperstein, J.J. Condon, and J.R. Boller, The Gamble I Pulsed Electron Beam Generator, J. Vac. Sci. Tech., 10: 961 (1973).

22. R.J. Commisso, Shyke A. Goldstein, R.A. Meger, J.M. Neri, W.F. Oliphant, P.F. Ottinger, F.C. Young, and B.V. Weber, Plasma Erosion Opening Switch Physics Investigations, Bull. Am. Phys. Soc., 28:1147 (1983).
23. B.V. Weber, R.J. Commisso, W.F. Oliphant, and P.F. Ottinger, Positive Polarity Operation of the Plasma Erosion Opening Switch, 1984 IEEE Conference on Plasma Science, St. Louis, MO (1984), IEEE Cat. No. 84CH1958-8.
24. J.M. Neri, R.J. Commisso, Shyke A. Goldstein, P.F. Ottinger, B.V. Weber, and F.C. Young, Plasma Opening Switch Studies, 1983 IEEE Conference on Plasma Science, San Diego, CA (1983), (IEEE Cat. No. 83CH1947-3).
25. J.M. Neri, R.J. Commisso, and R.A. Meger, Plasma Source Development for Plasma Opening Switches, Bull. Am. Phys. Soc., 27:1054 (1982).
26. R.J. Commisso, D.D. Hinshelwood, J.M. Neri, F.W. Oliphant, T.J. Renk, and B.V. Weber, Experimental Study of PEOS-Load Interaction, Bull. Am. Phys. Soc., 29:1206 (1984).
27. P.A. Miller, I.W. Poukey, and T.P. Wright, Electron Beam Generation in Plasma-Filled Diodes, Phys. Rev. Lett., 35:940 (1976).
28. P.F. Ottinger, S.A. Goldstein, and R.A. Meger, Theoretical Modeling of the Plasma Erosion Opening Switch for Inductive Storage Applications, J. Appl. Phys., 56:774 (1984).
29. D.D. Hinshelwood, private communication.
30. D.D. Hinshelwood, Explosive Emission Cathode Plasmas in Intense Relativistic Electron Beam Diodes, NRL Memorandum Report No. 5492 (1985); R.K. Parker, R.E. Anderson, and C.V. Duncan, Plasma-Induced Field Emission and the Characteristics of High-Current Relativistic Electron Flow, J. Appl. Phys., 45:2463 (1974)
31. I. Langmuir and K.B. Blodgett, Currents Limited by Space Charge Between Coaxial Cylinders, Phys. Rev. (Ser. 2), 22:347 (1923); I. Langmuir, The Effect of Space Charge and Residual Gases on Thermionic Currents in High Vacuum, Phys. Rev., (Ser. 2), 2:450 (1913).
32. B.V. Weber, R.J. Commisso, R.A. Meger, J.M. Neri, W.F. Oliphant, and P.F. Ottinger, Current Distribution in a Plasma Erosion Opening Switch, Appl. Phys. Lett., 45:1043 (1984).
33. B.G. Mendelev, Acceleration of the Deionization of a Rarefied Gas by Means of Back Voltage, J. Tech. Phys., (USSR), 21:710 (1951).
34. W. Koch, Experimenteller Nachweis Der Ionenschichten während der Entionisierung, J. Tech Phys., 17:446 (1936).
35. R.J. Barker and Shyke A. Goldstein, Computational Studies of a Radial Pinch-Reflex Diode, Bul. Am. Phys. Soc., 26:921 (1981).

36. Shyke A. Goldstein and R. Lee, Ion-Induced Pinch and the Enhancement of Ion Current by Pinced Electron Flow in Relativistic Diodes, Phys. Rev. Lett., 35:1079 (1975).
37. K.D. Bergeron, Two-Species Flow in Relativistic Diodes Near the Critical Field for Magnetic Insulation, Appl. Phys. Lett., 28:306 (1976).
38. P.F. Ottinger, J. Grossman, J. Neri, R. Kulsrud, D. Bacon, and A.T. Drobot, Theoretical Investigations of PEOS Operation, Bull. Am. Phys. Soc., 29:1342 (1984).
39. P.F. Ottinger, Operational Window for a PEOS Used for Voltage Multiplication on Pulsed Power Generators, NRL Memorandum Report 5591 (1985).
40. R.D. Ford, J.K. Burton, R.J. Commisso, G. Cooperstein, J.M. Grossmann, D.J. Jenkins, W.H. Lupton, J.M. Neri, P.F. Ottinger, and J.D. Shipman, Jr., Staged Inductive Pulse Generator with Capacitive Current Source, NRL Memorandum Report 5852, (1986).
41. J. Shannon, M. Wilkinson, R. Miller, and O. Cole, System Study for an Inductive Generator, Proc. Fifth IEEE Pulsed Power Conference, Washington, DC (1985), IEEE Cat. No. 85C 2121-2; J. Shannon, P. Krickhuhn, R. Dethlesen, O. Cole, and H. Kent, A Compact 1 MJ Capacitor Bank Module, Proc. Fifth IEEE Pulsed Power Conference, Washington, DC (1985), IEEE Cat. No. 85C 2121-2.

THE REFLEX SWITCH: A FAST-OPENING VACUUM SWITCH

Laszlo J. Demeter

Physics International Company

San Leandro, California 94577-0703

INTRODUCTION

   Pulsed power systems based on power amplification of magnetic energy stored in vacuum offer significant advantages in power scaling, compactness and cost over conventional technology. The key component of such systems is the vacuum switch, which is to stay closed for a long enough time to charge an inductor with current and to open on a fast time scale to produce a power amplified output pulse near the final vacuum load. The development of such a vacuum device, called the "reflex switch", is described herein. It is used to energize inductive stores with charging times of both tens and hundreds of nanoseconds. At opening, the stored magnetic energy is discharged on time scales compressed typically to 1/3 - 1/6 of the charging duration.

   In a conventional relativistic electron beam (REB) accelerator high-voltage, high-power pulses are frequently generated in liquid dielectric and injected through a dielectric-vacuum interface into a vacuum region where they are utilized to accelerate electron beams. At the insulator interface of the conventional approach a basic constraint exists. The power per unit area is limited by its electrical breakdown strength.[1] Due to this limitation high-power conventional systems are designed to have large-area insulators, and long, magnetically insulated vacuum transmission lines (MITL)[2] which are used to transport the energy to a central load.

   For sake of compactness and power scaling one would like to accumulate the energy in vacuum and then transfer this energy to the nearby load at a high power density. It has been recognized for some time that one way to accomplish this goal is to use magnetic energy storage in vacuum. The energy density in a magnetic field can be much greater than that of an electric field in dielectrics.

The basic elements of a typical magnetic energy storage system are shown in Fig. 1. The key component is an opening switch which can operate in vacuum. During the "closed" mode of this switch the generator builds up a current, I, in the storage inductor, L_s. When the current is near its peak value the switch makes a very fast transition to its "open" mode. This chokes the current and the resulting negative value of dI/dt causes the inductor to produce a large driving voltage across the parallel combination of the switch and load. The basic insulator problem is eliminated because energy is transported into the vacuum region through a low-voltage insulator at a relatively low power level over a long period of time. This energy is accumulated in a vacuum magnetic store and is subsequently converted into a short, high-power output pulse when the switch opens.

The principal difficulty with magnetic energy storage systems is to find a high-power opening switch which will operate in vacuum. One approach is to use plasmas driven by $\bar{J} \times \bar{B}$ forces.[3] Another is to apply the plasma erosion technique.[4-6] In the following sections a new type of high-power vacuum opening switch, the reflex switch, is described.

DESCRIPTION OF REFLEX SWITCH OPERATION

The operation of the reflex switch is based on reflex triode physics. As shown in Fig. 2, its main elements are a primary cathode, K_1, a thin anode, A, and a secondary cathode, K_2, which is electrically floating and reflects electrons back through the anode toward K_1. The reflexing electrons scatter and deposit their energy in the anode. An axial magnetic field is applied to prevent radial electron losses. Positive ions are accelerated from the anode to both cathodes.

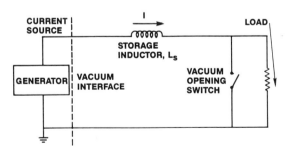

Fig. 1. Schematic of a vacuum magnetic energy storage system.

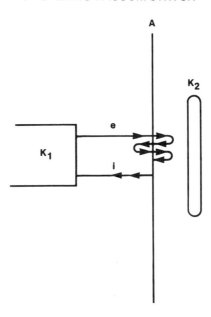

Fig. 2. Basic reflex switch configuration. K_1 is the primary cathode. A is a thin anode foil. K_2 is an electrically floating secondary cathode.

The combination of multiply reflected electrons and counter-streaming positive ions flowing between cathode, K_1, and anode, A, results in unique properties. One of the most important of these is that the total current can be orders of magnitude greater than the Langmuir-Child[7] (L-C) current appropriate for an ordinary vacuum diode of the same dimensions. Another characteristic is that during reflexing the triode acts as a constant-voltage device.

The reflex triode was originally investigated by Humphries et al.[8-10] The high-current mode of the reflex triode was predicted by Smith[11] from theoretical considerations. (Developments of the theoretical model are given in Ref's. 12 and 13). The first experimental observation of the high-current mode was made by Prono et al.[13] The properties of the theoretical model led Creedon[14] to suggest that the reflex triode could be converted into an opening switch by controlled termination of its electron reflexing. There is now a considerable amount of experimental evidence indicating that the basic properties of the theoretical model are correct.[13,15-17] Experiments with the reflex triode used as a switch have provided further confirmation of the theory and have demonstrated the opening phase of the reflex switch.[18,19,20]

The high-current mode of the reflex triode is associated with the formation of a narrow anode sheath, across which most of the AK_1 potential difference occurs. A detailed description of the sheath formation process can be found in Ref's. 12 and 13. Only a brief discussion is given here.

The form of the potential distribution in the reflex triode is shown in Fig. 3. Define η as the average number of times an electron encounters the anode prior to stopping in it. In the limit of very low voltage across the $A-K_1$ gap, $\eta=1$, and the potential distribution is that of ordinary Langmuir bipolar flow. The current is about 1.9 times the L-C value. The value of η increases with the $A-K_1$ gap potential, V. For $\eta>1$, the potential distribution is distorted by the space charge of the multiply reflected electrons and the corresponding positive ion flow. The greater the value of η, the greater is this distortion, until η approaches a critical number, η_c, corresponding to a critical gap potential, V_R. For $\eta=\eta_c$ the potential distribution is relatively flat in most of the $A-K_1$ gap, and most of the potential difference is concentrated in a thin "sheath" near the anode, as shown in Fig. 3. This thin sheath becomes the effective diode gap instead of the much larger $A-K_1$ gap. It is due to this small effective gap that the reflex mode gives very high-current operation relative to an ordinary Langmuir bipolar flow.

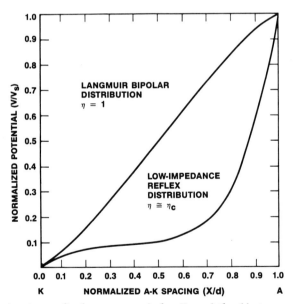

Fig. 3. Variation of the potential, V, with distance, X, in a reflex triode, V_s is the anode potential and d is the A-K spacing. η is the average number of electron transits through the anode foil.

As V approaches V_R and η approaches η_c, the gap operates essentially as a constant-voltage device with $V=V_R$ over a large range of current. (This means that with increasing current density the width of the anode sheath should become correspondingly smaller). The value of V_R depends on the energy spectrum of reflexing electrons and on the anode thickness. The energy spectrum determines η_c which, in turn, determines the V_R value required to give electrons enough energy to pass through a given anode η_c times. For a given electron energy spectrum, increasing the anode thickness raises V_R.

The voltage-current characteristics of a reflex switch are shown in Fig. 4. Figure 4a is a plot of switch current versus switch voltage; Fig. 4b is a plot of switch voltage versus time. The three dashed curves in Fig. 4a represent the characteristics of the device in various modes of operation. The heavy curve in Fig. 4a shows the time-dependent path of the switch in the current-voltage plane.

The curve marked I_R represents the low-impedance reflexing mode at the potential, V_R. The curve marked I_{BP} is the ordinary bipolar mode[21] with electrons from K_1 and ions from A but no reflexing electrons. The curve marked I_{ER} corresponds to the presence of reflexing electrons but no ions.

The time dependence of the switch voltage shown in Fig. 4b consists of four phases. (These phases are also marked on the heavy curve of Fig. 4a.) In Phase 1 only reflexing electrons are present in the A-K_1 gap. The switch follows the characteristic curve I_{ER} and the current is low. When an anode plasma forms, ions begin to flow from the anode and the switch makes a transition to Phase 2. In Phase, 2, the switch current operates on the reflex triode characteristic curve, I_R, at the voltage, V_R. This is the low-impedance, constant voltage, "closed" mode. The spacing between K_2 and A (see Fig. 2) is picked so that plasma motion or other processes will cause K_2 to short to A when the current has built up to the desired value. When this shorting occurs, the electrons from K_1 stop in K_2 and reflexing ends. The operation enters Phase 3, which is a fast transition from I_R toward the ordinary bipolar characteristic, I_{BP}. This is the "opening" phase of the switch during which the voltage increases rapidly to a peak, V_o, caused by the discharge of the inductor. The current-carrying electrons are accelerated up to kinetic energies, $E_K = eV_o$, and impinge on K_2 which now is part of the anode. Thus, in Phase 3 the reflex switch becomes a source of powerful electron beams and radiation, if K_2 is chosen to be a bremsstrahlung converter. In Phase 4 the switch has fully "opened" to the Langmuir bipolar characteristic corresponding to the physical geometry of the A-K_1 gap and the pulse decays along the I_{BP} curve.

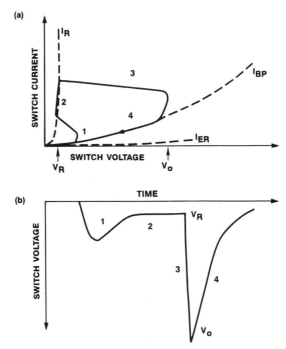

Fig. 4. Current and voltage behavior during reflex switch operation. (a) shows switch current vs. switch voltage. The curves I_{ER}, I_R and I_{BP} represent the voltage-current characteristics of the switch in different modes of operation. The solid curve represents the path in the IV plane. (b) is a plot of switch voltage vs. time. Four different phases of operation are identified by the numbers next to this curve.

The opening of a reflex switch is a very complex process and is not fully understood at present. However, certain basic processes must occur. In general, the distribution of the electrons can adjust much more rapidly than that of the ions. Thus when K_2 shorts to A and electrons are no longer reflected, the excess electrons in the primary gap should leave on a subnanosecond time scale unless they are trapped by the electric field.

During the closed phase of the switch the positive ion current is much greater than for bipolar flow[12,13]. In the A-K_2 gap these ions have mainly been accelerated by the potential difference ΔV, of the anode "sheath". (ΔV is ~ 0.9 V_R for the example of Fig. 3.) The switch cannot reach the bipolar mode until these high density ions leave the A-K_1 gap. Thus, the opening time for the switch should be determined by the time required to reduce the ion charge density existing in the reflexing phase to that of a bipolar flow.

THE REFLEX SWITCH: A FAST-OPENING VACUUM SWITCH

EXPERIMENTS AND RESULTS

Some early reflex switch experiments were conducted on two pulse generators at Physics International: the 225-W[22] (1 MV, 5.25 Ω, 65 ns) and the OWL II[23] (1 MV, 2 Ω, 120 ns). A typical experimental setup depicting the 225-W (CAMEL) generator as the current source is shown in Fig. 5. The size of the coaxial, vacuum storage inductor used for these experiments was typically 160 nH. In the high-current OWL II tests it was occasionally increased to 300 nH. The anode foil, A, was a clear polypropylene plastic. The primary cathode, K_1, was typically either an annular-rim type or one using an annular array of roll pins. The electrically floating secondary cathode, K_2, was a stainless steel or plastic disc mounted on a cantilevered nylon rod. Both the primary diode gap, A-K_1 and the secondary gap, A-K_2, were adjustable. The dimensions of K_1, K_2, and the diode gaps ranged from 1 to 10 cm. The pressures maintained in the reflex switch were typically in the low-10^{-4} or high-10^{-5} Torr range. An axial magnetic guide field of variable strength up to 3 T was generated in the K_1/A/K_2 region by a pair of Helmholz coils exterior to the stainless steel vacuum chambers and centered in the anode plane. Current monitors placed on both primary and secondary sides of the anode were used to follow the temporal history of magnetic energy accumulation and discharge and to check on possible current losses. The primary diode (switch) voltage was defined by electronically correcting the generator output voltage, V_G, for the inductive voltage drop between the generator and K_1. A scintillator/photo-diode combination was used to monitor the bremsstrahlung radiation generated when high-energy electrons impinged on K_2.

The signal traces obtained from an experiment with the CAMEL generator are shown in Fig. 6. The peak voltage during Phase 1 is 560 kV. When the anode plasma was formed (by self-flashover of the anode foil) the low impedance mode was established in about 20 ns. The reflexing voltage, V_R, is 100 kV in Phase 2. The peak open-switch voltage, V_o, in Phase 3 was 1.8 MV. The 10%-90% risetime

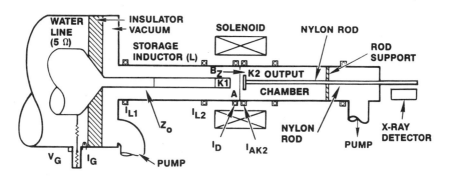

Fig. 5. Apparatus used in the experimental program.

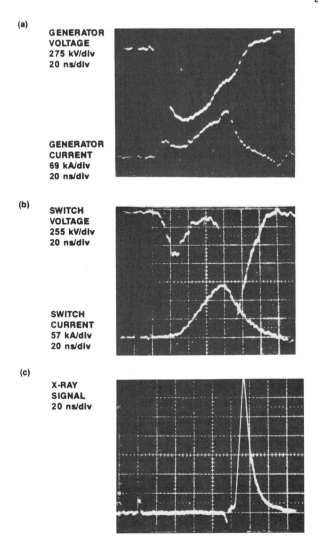

Fig. 6. Results of an experiment with the CAMEL generator. (a) shows the generator voltage and current. The switch voltage and current are shown in (b). The signal from the x-ray detector is shown in (c).

was ~ 12 ns. Switch impedance increased at a rate in excess of 1 ohm/ns during the voltage risetime. (The net impedance transition was from 0.8 ohm just prior to switching to 13 ohms at peak voltage.) The switch current in Fig. 6 increased to a value of

160 kA during the 48 ns duration of Phase 2, and was 140 kA at the time of peak accelerating voltage. During the low-impedance phase of the switch, the inductive store accumulated energy at an average rate of 65 GW. Peak energy in the inductor was 3.4 kJ. The magnetic energy of the store was converted to REB kinetic energy with essentially 100% efficiency. The x-ray pulse occuring at the same time as the high-voltage pulse across the switch. Its risetime being 8 ns and the pulse duration 15 ns (FWHM). The constant voltage nature of the reflex triode is clearly evident from the data shown in Fig. 6. (During the closed mode, the switch current increased by a factor of two while the reflexing voltage stayed essentially constant.)

The opening of the switch must be initiated by an impedance collapse in the A-K_2 gap. Consequently the length of time that the switch spends in the low-impedance mode increases with A-K_2 spacing. If the A-K_2 gap is sufficiently large, the switch will not open during the generator pulse.

During the reflexing mode, electrons can return to A by two processes. Either the electrons are reflected by the potential distribution between A and K_2 or they strike K_2 and are replaced by electrons emitted from K_2. Measurement of K_2 potential and the damage pattern observed on K_2 indicate the latter to be the typical case. Also, a zero reading on I_{AK2} (see Fig. 5) implies that the current of electrons striking K_2 must be equal to the sum of the electron current emitted from K_2 and the ion current flowing from A to K_2.

Figure 7 shows an example of the experimental results obtained with the OWL generator at higher current levels. The peak current shown is 340 kA. The peak voltage in Phase 1 being 800 kV. The reflexing voltage, V_R in the closed mode is 300 kV. The peak switch voltage in the opening phase is 2.7 MV. Typically, the diode voltage dropped to zero after the main high-voltage pulse.

Figure 8 displays the measured voltage-current characteristic for the CAMEL experimental data of Fig. 6. (This can be compared with the theoretical curve in Fig. 4a.) The numbers next to this curve represent the time in nanoseconds measured from the beginning of the current pulse. As discussed before, the diode current after opening (Phase 4 in Fig. 4a) is expected to be Langmuir bipolar flow, I_{BP}.[24] Comparing the measured current to that computed for the original diode dimensions indicates that the primary gap is significantly reduced during the operation of the switch. By invoking plasma closure mechanism as responsible for the change one can infer the average "closure" velocity being in excess of 30 cm/μs, which is an order of magnitude greater than that of the usual anode and cathode plasmas observed in conventional diodes.

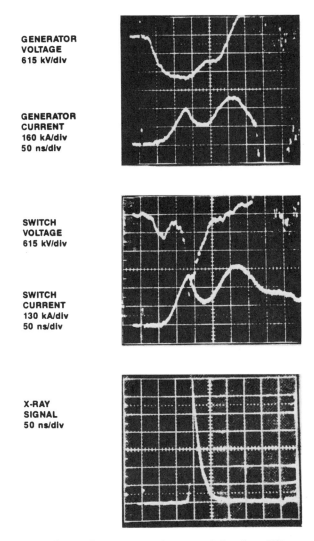

Fig. 7. Results of an experiment with the OWL generator.

Anomalous impedance collapse in a reflex triode has been observed by others. Prono et al.[16,17] have measured equivalent "velocities" of 10-20 cm/μs and shown that the time from current turn-on until the short occurs is a non-linear function of the A-K spacing. They have developed a theory showing that high-velocity neutral atoms can be produced in the anode plasma by a charge-exchange process[17]. If these neutrals are reionized in the A-K_1 gap the diode current could be much greater than the expected bipolar flow. The existence of rapidly moving neutrals in other diode experiments has also been observed[25].

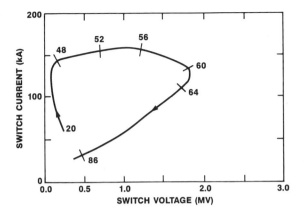

Fig. 8. Voltage-current plot for the experiment of Fig. 6. The numbers next to the curve are the times in nanoseconds measured from the beginning of the current pulse.

In any case, the fast gap-shorting process seems to be a fact of life in the operation of the reflex switch. Its mechanism is not fully understood at present. It could be due to ionized neutrals or some fast-expanding conducting region(s) in the gap.

As discussed before, the impedance collapse in the A-K_2 gap triggers the opening of the switch. This impedance collapse also occurs in an anomalously short time. This phenomenon probably involves the same mechanisms which are at work in the A-K_1 gap.

SUMMARY

The reflex switch has been demonstrated to be a fast, high-power vacuum opening switch. It was successfully used to charge vacuum inductors with high currents injected at low power level through a conventional interface insulator. When it opened, the stored magnetic energy was discharged as a fast-risetime, power-amplified output pulse, which was utilized to generate intense, relativistic electron beams. It was experimentally used in conjunction with magnetic energy storage over a wide range of currents and power-levels.

ACKNOWLEDGMENTS

The author wishes to thank the all-important contribution of J. Creedon. In addition to being the originator of the opening-reflex-triode idea, he was most helpful to the experiments by providing theoretical support and analysis. The unique reflex triode experience of D. Prono was a most welcome complement to our team effort. In furnishing useful ideas and calculations the roles of S. Humphries, V. Bailey, and R. Genuario are gratefully acknowledged. The success of the reflex switch experiments could not have been achieved without the individual contributions of B. Ecker, S. Glidden, H. Helava, G. Proulx, C. Eichenberger, G. James, D. Drury, L. Sanders, T. D'Agnostino, and S. Love.

This work was supported by the Defense Nuclear Agency.

REFERENCES

1. L.I. Rudakov, Transport of a Relativistic Electron Beam to a Fusion Target, Sov. J. Plasma Phys., 4(1):40 (1978).
2. M.S. Di Capua, Magnetic Insulation, IEEE Trans. Plasma Science, PS11:205 (1983).
3. L.J. Demeter and V.L. Bailey, "Energy Storage Compression and Switching," V. Nardi, H. Shalin, and W.H. Bostick, ed. Plenum Press, New York (1983).
4. P.A. Miller, J.W. Poukey, and T.P. Wright, Electron Beam Generation in Plasma-Field Diodes, Phys. Rev. Letters, 35:940 (1975).
5. C.W. Mendel, Jr. and S.A. Goldstein A Fast-Opening Switch for Use in REB Diode Experiments, J. Appl. Phys., 48:1004 (1977).
6. R.A. Meger, R.J. Commisso, G. Cooperstein, and S.A. Goldstein, Vacuum Inductive Store/Pulse Compression Experiments on a High Power Accelerator Using Plasma Opening Switches, Appl. Phys. Lett., 42:943 (1983).
7. I. Langmuir, The Effect of Space Charge and Residual Gases on Thermionic Currents in High Vacuum, Phys. Rev., 2:450 (1913).
8. S. Humphries, J.J. Lee, and R.N. Sudan, Generation of Intense Pulsed Ion Beams, Appl. Phys. Letters, 25:20 (1974).
9. S. Humphries, Jr., J.J. Lee, and R.N. Sudan, Advances in the Efficient Generation of Intense Pulsed Proton Beams, J. Appl. Phys., 46:187 (1975).
10. S. Humphries, Jr., R.N. Sudan, and W.C. Condit, Jr., Production of Intense Megavolt Ion Beams with a Vacuum Reflex Discharge, Appl. Phys. Lett., 26:667 (1975).
11. Ian Smith (Private Communication with J. Creedon).
12. J.M. Creedon, I.D. Smith and D.S. Prono, Method of Generating Very Intense Positive-Ion Beams, Phys. Rev. Lett., 35:91 (1975).
13. D.S. Prono, J.M. Creedon, I. Smith, and N. Bergstrom, Multiple Reflections of Electrons and the Possibility of Intense Positive-Ion Flow in High ν/γ Diodes. J. Appl. Phys., 46:3310 (1975).

14. J. Creedon, S. Putnam, and I. Smith, US Patent No. 4080549 (1978).
15. D.S. Prono, J.W. Shearer, and R.J. Briggs, Pulsed Ion Diode Experiment, Phys. Rev. Lett., 37:21 (1976).
16. D.S. Prono, H. Ishizuka, B. Stallard, and W.C. Turner, LLL's Long Pulse Reflex Diode Ion Source (IPINS), Bull. Amer. Phys. Soc., 23:903 (Sept. 1978).
17. D.S. Prono, H. Ishizuka, E.P. Lee, B.W. Stallard, and W.C. Turner, Charge-Exchange Neutral-Atom Filling of Ion Diodes; Its Effect on Diode Performance and A-K Shorting, J. Appl. Phys., 52:3004 (1981).
18. L. Demeter, B. Ecker, C. Eichenberger, H. Helava, T. Naff, G. Proulx, and J. Creedon, Fast Discharge of a Vacuum Magnetic Energy Store Using the Reflex Switch, Bull. Amer. Phys. Soc., 27:1030 (1982).
19. B. Ecker, J. Creedon, L. Demeter, S. Glidden, and G. Proulx, The Reflex Switch: A High-Current, Fast-Opening Vacuum Switch, 4th IEEE Pulsed Power Conf., Albuquerque, NM, June 6-8, 1983, IEEE Cat. No. 83CH1908-3.
20. J. Creedon, L.J Demeter, B. Ecker, C. Eichenberger, S. Glidden, H. Helava, and G.A. Proulx, The Reflex Switch - A New Vacuum Opening Switch for Use with Magnetic Energy Storage, J. Appl. Phys., 57:1582 (1985).
21. I. Langmuir, The Interaction of Electron and Positive Ion Space Charges in Cathode Sheaths, Phys. Rev., 33:954 (1929).
22. Model 225-W Pulserad, Operation and Maintenance Manual, Report No. PIMM-1568, Physics International Co., San Leandro, CA (1975).
23. G.B. Frazier, OWL II Pulsed Electron Beam Generator, J. Vac. Sci. Technol., 12:1183 (1975).
24. J.W. Poukey, Ion Effects in Relativistic Diodes, Appl. Phys. Lett., 26:145 (1975).
25. R. Pal and D. Hammer, Anode Plasma Density Measurements in a Magnetically Insulated Diode, Phys. Rev. Lett., 50:732 (1983).

PLASMADYNAMIC OPENING SWITCH TECHNIQUES

P. J. Turchi

RDA Washington Research Laboratory
Alexandria, VA 22314

INTRODUCTION

Circuit interruption techniques that depend on plasma dynamics comprise a number of quite different physical phenomena ranging from plasma discharge quenching in the arc chutes of power line breakers to non-neutral plasma-sheath processes near electrodes. The present discussion examines some aspects of electromagnetic power flow in the context of dynamic plasma switching techniques. Such techniques involve the control of conducting-mass distributions in order to accumulate energy at low power and release it at high power. A rather general formulation is developed and the Plasma Flow Switch is used as an example of a particular concept. The intent is to establish a reasonable basis for considering circuit interruption concepts that involve dynamic plasmas, either inadvertently or by design.

BASIC FORMULATION FOR POWER FLOW WITH PLASMA ELEMENTS

The principal concern in a pulsed power system is power flow from an energy source (e.g., capacitor bank or inductive storage coil) to a load (e.g., diode, flashlamp) through a number of intermediate elements comprising passive components, such as transmission lines and insulators, and components such as switches, in which some action is prescribed. Maxwell's equations, along with relations specifying material properties, will govern the behavior of the various components and the interactions between components. A useful distillation of these equations for examining power flow is Poynting's theorem in differential form:

$$\frac{\partial}{\partial t}\left[\frac{\epsilon E^2}{2} + \frac{B^2}{2\mu}\right] = -\bar{E}\cdot\bar{j} - \nabla\cdot\bar{S}, \qquad (1)$$

where $\bar{S} = \bar{E}\times\bar{B}/\mu$ is the appropriately-named Poynting vector and j is the current density. (Note that the equation used here is a simplified form assuming linear behavior of the material, i.e., $\bar{D} = \epsilon\bar{E}$ and $\bar{B} = \mu\bar{H}$).

For a simple electrical circuit consisting of a voltage source and a resistor, the first term on the right-hand side is the resistive dissipation (per unit volume) associated with Ohm's law ($\bar{E} = \eta\bar{j}$, where η = resistivity). The left-hand side of the equation thus represents the loss of electromagnetic energy from a volume due to dissipation within that volume. The second term on the right is then identified as a rate of loss of energy from a volume because of a net difference in flux, $\bar{S} = \bar{E}\times\bar{B}/\mu$, through the surface surrounding that volume. The Poynting vector thereby specifies the local electromagnetic power flux. Flow of electromagnetic energy in a pulsed power system can therefore be evaluated in terms of the local electric and magnetic fields existing at interfaces between components (or at stations within a component, e.g., along a transmission line). For many components, the electric field is obtained from the spatial distributions of electric charge (i.e., by solution of Poisson's equation). In switching elements that involve plasmas, however, the mobility of charges on the time-scale of interest may impose a situation of quasi-neutrality, and the electric field is more usefully derived from the self-consistent dynamics of the charge-carriers. Such dynamics are usually embodied in a so-called generalized Ohm's law.

There are two basic ways of constructing a generalized Ohm's law. The dynamics of the charges may be calculated from their momentum equations, or the electric field in a frame at rest in the laboratory can be assembled by adding various classical effects (Ohm's law, Hall effect, Thomson effect, etc.) and then be transformed to a moving frame of reference. The latter is more general (if all effects are empirically known). The former is more physically transparent, however, (if simple-dynamics are assumed) and will therefore be used here.

For a quasi-neutral plasma ($n_e = n_i$, $Z = 1$), the fluid momentum equation for the electrons may be written as:

$$n_e m_e \frac{D_e \bar{u}_e}{Dt} = -n_e e(\bar{E} + \bar{u}_e \times \bar{B}) - \nabla p_e - n_e m_e \nu_{ei}(\bar{u}_e - \bar{u}_i) \qquad (2)$$

where $D_e/Dt = \partial/\partial t + u_e \cdot \nabla$ is the convective derivative based on the electron fluid velocity $\bar{u}_e$, p_e is the (isotropic) electron pressure, $\bar{u}_i$ is the ion fluid velocity, and ν_{ei} is the electron-ion collision frequency for momentum transfer. In this simple case, the difference in electron and ion fluid velocities may be written in terms of the local current density:

$$\bar{u}_e - \bar{u}_i = -\frac{\bar{j}}{n_e e} \ . \tag{3}$$

If the electron inertia is neglected on the temporal- and spatial-scale of interest, then the left-hand side of the momentum equation is set to zero, and the electric field is obtained consistent with motion of the electron fluid in the presence of magnetic fields, pressure gradients, and collisions:

$$\bar{E} = \left(\frac{m_e \nu_{ei}}{n_e e^2} \right) \bar{j} - \bar{u}_e \times \bar{B} - \frac{\nabla p_e}{n_e e} \ . \tag{4}$$

The term in parentheses may be identified as the electrical resistivity:

$$\eta = \frac{m_e \nu_{ei}}{n_e e^2} \ . \tag{5}$$

If the electron fluid velocity is eliminated in favor of the heavy particle (ion) fluid velocity $\bar{u}$ and the current density, the generalized Ohm's law may be written as:

$$\bar{E} = \eta \bar{j} - \bar{u} \times \bar{B} + \frac{\bar{j} \times \bar{B}}{n_e e} - \frac{\nabla p_e}{n_e e} \ . \tag{6}$$

The four terms on the right may then be respectively identified with the classical Ohm's law, the transformation for conductor motion in a magnetic field, the Hall effect in a non-moving conductor, and a term that can lead to the Thomson effect (electric field due to temperature gradient).

If for simplicity of discussion the last term is ignored, then substitution of the generalized Ohm's law for the electric field in the Poynting vector gives:

$$\bar{S} = \left(\frac{\eta}{\mu} \right) \bar{j} \times \bar{B} - \frac{(\bar{u} \times \bar{B}) \times \bar{B}}{\mu} + \frac{(\bar{j} \times \bar{B})}{n_e e} \times \frac{\bar{B}}{\mu} \ . \tag{7}$$

The first term on the right is associated with power flow due to resistive diffusion at a conducting boundary surface, and points in the direction of the Lorentz force ($\bar{j} \times \bar{B}$). Inductor-to-inductor energy transfer using a resistor as the opening switch utilizes this first term. Energy is retained in an inductive store by electrical conductors (low η). A switch region (high η) acts as a gate at the boundary of an inductive store, releasing energy to another circuit.

For the second and third terms on the right-hand side, expansion of the vector product provides:

$$-\frac{(\bar{u} \times \bar{B}) \times \bar{B}}{\mu} + \left(\frac{\bar{j} \times \bar{B}}{n_e e}\right) \times \frac{\bar{B}}{\mu} = \left(\bar{u}_\perp - \frac{\bar{j}_\perp}{n_e e}\right) \frac{B^2}{\mu} , \qquad (8)$$

where the subscript '$\perp$' indicates the component perpendicular to the magnetic field vector. If the relative drift velocity of the charge carriers, $\bar{j}_\perp/n_e e$, is low compared to the conductor velocity, $\bar{u}_\perp$, then the power flow through an area, A, at the boundary is:

$$\bar{S} \cdot \bar{A} = \frac{\eta}{\mu} (\bar{j} \times \bar{B}) \cdot \bar{A} + \left[\bar{u}_\perp \cdot \bar{A} \left(\frac{B^2}{2\mu}\right)\right] + \left[\bar{u}_\perp \cdot \bar{A} \left(\frac{B^2}{2\mu}\right)\right] , \qquad (9)$$

where the last two equal terms were separated from Eq. (8) to display, respectively, the rate of work performed by magnetic pressure, $B^2/2\mu$, acting through a rate of volume change $\bar{u}_\perp \cdot \bar{A}$, and the rate at which magnetic energy, at density $B^2/2\mu$, is supplied to a new volume that is created at a rate $\bar{u} \cdot \bar{A}$. The first term on the right is due to resistive diffusion at a conducting boundary.

For a perfect gas, the energy density, w_d, and the pressure, p, are related by the specific heat ratio, γ:

$$w_d = \frac{p}{\gamma-1} , \qquad (10)$$

Identification of the magnetic energy density and the magnetic pressure thus leads to the notion that (for displacements perpendicular to the magnetic field) magnetic energy in a volume, V, behaves like that of a perfect gas with $\gamma = 2$. If the usual isentropic relationship is employed, pV^γ = constant, the magnetic pressure thus varies inversely as the square of the area (perpendicular to the field), a condition corresponding to conservation of magnetic flux ($\bar{B} \cdot \bar{A}$). The non-conservative effects in the Poynting vector flux are associated with the resistive diffusion term, $(\eta/\mu)\bar{j} \times \bar{B}$. Magnetic flux may still be conserved (like the particle number in a perfect gas), but diffusion results in the loss of energy irreversibly to heat.

The relative importance of convection of magnetic energy vs resistive diffusion may be estimated by combining the generalized Ohm's law with Faraday's law:

$$\bar{E} = \eta \bar{j} - \bar{u} \times \bar{B} \qquad (11)$$

and

$$\nabla \times \bar{E} = - \frac{\partial \bar{B}}{\partial t} \qquad (12)$$

(where pressure gradient terms have been neglected and $\bar{u}_e \approx \bar{u}$ for simplicity). After some straightforward application of vector identities, Ampere's law and conservation of mass, the following equation is obtained: (For uniform conductivity and rectangular flow)

$$\frac{D}{Dt}\left(\frac{\bar{B}}{\rho}\right) = \frac{\nabla^2 \bar{B}}{\sigma \mu \rho} \qquad (13)$$

where $D/Dt = (\partial/\partial t + \bar{u} \cdot \nabla)$ is the convective derivative based on the fluid velocity $\bar{u}$, ρ is the mass density, and a uniform conductivity ($\sigma = 1/\eta$) is assumed. If dimensionless variables are defined as:

$$\alpha = x/x_o, \quad \omega = u/u_o, \quad \theta = u_o t/x_o$$

$$\beta = B/B_o, \quad \text{and} \quad \zeta = \rho/\rho_o$$

where the subscript 'o' denotes characteristic values (e.g., x_o is the characteristic spatial dimension), then a normalized differential equation is obtained:

$$\frac{D}{D\theta}(\bar{\beta}/\zeta) = \frac{1}{\sigma \mu u_o x_o} \frac{1}{\zeta} \nabla^2 \bar{\beta} \qquad (14)$$

The collection of terms $R_m = \sigma \mu u_o x_o$ is the magnetic Reynolds number. For $R_m \gg 1$, the convective derivative of $(\bar{\beta}/\zeta)$ is nearly zero, so the quantity $(\bar{\beta}/\rho)$ is maintained in each parcel of conducting fluid as it travels through the flow-field. The close coupling of the magnetic field to the fluid is the so-called "frozen-field" approximation. As a fluid parcel expands or contracts, its mass (with volume per unit depth, a) is ρa = constant, so the magnetic flux $\bar{B} \cdot \bar{a}$ is constant. (For conducting boundaries surrounding flux in a nonconducting region, the enclosed flux will be nearly constant if the magnetic Reynolds number associated with the conducting material is very high. Care must be taken in defining x_o in this case).

Low magnetic Reynolds number ($R_m \ll 1$) corresponds to the fully diffusive case. The magnetic field distribution will depend on the conductivity distribution and boundary conditions, (e.g., $\nabla^2 B = 0$) for uniform conductivity). Magnetic flux can dif-fuse through the conductor as current is diverted from one current path to another. The magnetic flux contained by the conductor will thus change with time. The total magnetic flux in the system, however, may remain constant (depending on the circuit).

In both the high and low R_m limits, magnetic energy contained by the conductor can change. For the low R_m case, resistive diffusion is associated with energy lost to resistive heating. The high R_m situation allows energy to be interchanged reversibly between magnetic energy and work (or kinetic energy) as the conductor changes size and shape.

Opening switch techniques that involve plasmas must operate in context of one or more of the terms in the Poynting vector equation using Eqs. (7) and (8):

$$\bar{S} = \left(\frac{\eta}{\mu}\right) \bar{j} \times \bar{B} + \left(\bar{u}_\perp - \frac{\bar{j}_\perp}{n_e e}\right) \frac{B^2}{\mu} , \qquad (15)$$

where the low and high magnetic Reynolds number limits are represented by the first and second set of terms, respectively. Classical breakers and fuses utilize quenching media, such as high pressure SF_6 or sand, to increase the resistivity in a switch region, thereby allowing rapid diffusion of magnetic flux. Diffuse discharge switches also belong to this group. Dynamic plasma systems, such as plasma flow switching coupled to an implosion load, involve convection of magnetic flux and a sequence of exchanges between electromagnetic and kinetic energies. Other switch concepts, such as the plasma erosion switch and the magnetoplasmadynamic switch are based on the loss of sufficient charge-carriers to handle the local current. Microinstabilities and sheath growth occur. The former result in higher values of resistivity and are thus associated with increased resistive diffusion, while the latter can lead to magnetically-insulated electron flow, thereby working with the Hall term $\bar{j}_\perp/n_e e$.

Since techniques based on the resistive diffusion term (e.g., breakers, fuses) are discussed elsewhere in this book, the present discussion will focus on use of the convective terms. The high magnetic Reynolds number approximation will therefore be used in estimating plasma and magnetic field distributions and assessing the Poynting vector power flux.

PLASMADYNAMIC OPENING SWITCH TECHNIQUES

PLASMA CONVECTION OF MAGNETIC ENERGY

For switching concepts involving dynamic plasmas, the accumulation and subsequent delivery of magnetic energy depends on the evolution of plasma and magnetic field distributions[1,2,3]. Typically, (Fig. 1) an upstream vacuum region storing magnetic energy is separated from a downstream field-free vacuum region by a conducting element that is subject to electromagnetic forces and electrical heating. The conducting element prevents energy transfer to the downstream load as magnetic energy is being stored in the upstream region. The element then moves to allow rapid energy transfer to the downstream region. The conducting element may be at various times and in various embodiments a solid, liquid, vapor or plasma. Non-moving boundaries near the element may also be in various states. The general problem is quite complex, but amenable, in principle, to numerical calculation using state-of-the-art computer codes. In the following discussion, analytical techniques are applied to idealized cases in order to achieve some basic understanding of problems associated with dynamic plasma opening switch techniques.

For typical operation, such as the plasma flow switch, plasma conductivities are σ = 5-10 kS, u_o = 7-10x10^4 m/s, and x_o = 2-3 cm, so $R_m \approx 10$. At this level of magnetic Reynolds number, it is reasonable to adopt the frozen-field approximation which introduces two critical points for dynamic plasma opening switch technology: 1) plasma motion can convect magnetic field and therefore current and 2) plasma inertia can limit the rate at which magnetic energy is convected. The first point is important because the operation of a dynamic switch may depend on features such as the magnetic field difference across a moving and/or vaporizing element. Also, the downstream electrical and mechanical environ-

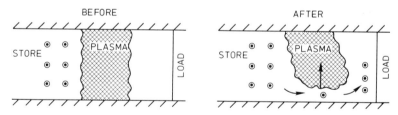

Fig. 1. A region of high electrical conductivity (e.g., plasma) can prevent magnetic energy from reaching a load. Opening switch action is achieved by creating a gap in the conducting material (e.g., by plasma motion) thereby releasing magnetic energy to the load. Since the conductivity of a plasma may depend only weakly on density, flow of energy to the load can be limited by residual plasma in the gap.

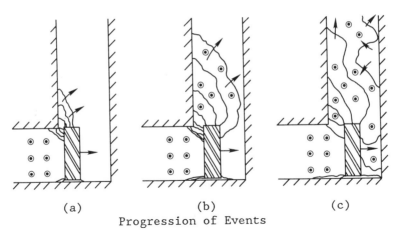

Fig. 2. As a gap opens between an electrode and a moving solid or liquid switch element, plasma will be accelerated carrying magnetic field and current downstream toward the load. Magnetic energy and pressure can thereby reach the load before the metal-to-metal gap has increased sufficiently to withstand high voltages.

ment may be degraded by a current "precursor" due to early plasma convection (Fig. 2a). The second point is critical for the power flow required through the switch region after dynamic operation of the switch has been completed. The presence of plasma generated off electrodes, insulators and contacts or left behind by the main switching element reduces the value of $u_\perp$ and thus limits the power delivered (Fig. 2c).

To examine the distribution of plasma mass during switch operation, it is necessary to specify the flow geometry. As a reasonably general first step, consider the motion of a magnetized plasma as it expands into a field-free vacuum. Such expansion will occur as metal-to-metal contact is broken for a magnetic gate switch or other moving solid-armature switches, and also for plasma armatures formed by vaporization or injection of material between two electrodes at high voltage. For the idealized case of a semi-infinite plasma of initially uniform density and field upstream of the vacuum region (Fig. 3), analytic solution is possible using the centered-expansion fan technique of one-dimensional, unsteady gas dynamics.

Basically, as a piston is removed from a long tube filled with stationary gas of sound speed a_0 and specific heat ratio γ, a continuous sequence of waves may be envisioned that propagate into the gas, causing motion of the gas to follow the piston. The first wave travels at the sound speed a_0. The second wave travels more slowly in the laboratory frame since the gas is now moving toward

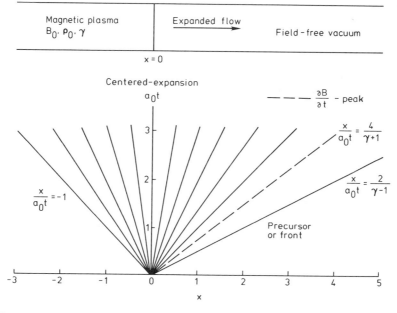

Fig. 3. An idealized representation of magnetic energy transfer in the presence of high conductivity plasma consists of a semi-infinite, uniform mixture of plasma and magnetic field ($-\infty < x < 0$) suddenly adjacent to a field-free vacuum region ($x>0$). The plasma expands adiabatically, with a specific heat ratio, γ, resulting in a region of nonuniform density and magnetic field between a characteristic line propagating upstream at $-a_0 t$ and one moving downstream at $2a_0 t/(\gamma-1)$, where a_0 is the sound speed in the initial plasma/field mixture. The nonuniform magnetic field region corresponds to a plasma discharge that spreads out with time.

the piston and has been expanded slightly to a lower sound speed value by the passage of the first wave. Subsequent waves move even more slowly as the gas flow to follow the piston is established. In fact, some later waves are carried (in the laboratory frame again) in the direction of the piston motion by the flow created by the previous waves. If the piston is removed very rapidly, a vacuum region can occur between the piston and the last wave, corresponding to complete expansion of the gas to zero density. This wave travels in the laboratory frame at the limiting speed $2a_0/(\gamma-1)$. Sudden expansion of a gas into vacuum is equivalent to extraction of a piston at speeds in excess of the limiting speed for the gas.

For an initially uniform, perfect gas, the density in the flow depends only on the local flow speed, u:

$$\frac{\rho}{\rho_o} = \left[1 - \frac{(\gamma-1)}{2}\frac{u}{a_o}\right]^{2/(\gamma-1)} \quad (16)$$

where ρ_o is the initial mass density. The flow speed, however, is related to the position x of the wave which provides this flow speed at time t:

$$\frac{x}{t} = \frac{\gamma+1}{2} u - a_o \quad (17)$$

(The waves or characteristic lines comprising the centered-expansion fan are shown in Fig. 3). Thus, the density distribution at any time is given by:

$$\frac{\rho}{\rho_o} = \left[1 - \left(\frac{\gamma-1}{\gamma+1}\right)\left(1 + \frac{x}{a_o t}\right)\right]^{2/(\gamma-1)} \quad (18)$$

Suppose the gas is a good electrical conductor so that the frozen-field approximation is valid. Specification of the initial distribution of magnetic field would then allow calculation of the magnetic field distribution as the gas expands (as long as the expansion is still characterized by the specific heat ratio γ). For the particular simple case of uniform magnetic field upstream of the vacuum ($B = B_o$, $\rho = \rho_o$ for $-\infty < x < 0$, $t=0$ and $\rho/\rho_o = B/B_o$, the magnetic field distribution at later times is:

$$\frac{B}{B_o} = \left[1 - \left(\frac{\gamma-1}{\gamma+1}\right)\left(1 + \frac{x}{a_o t}\right)\right]^{2/(\gamma-1)} \quad (19)$$

Note that at the origin (x=0) the relative magnetic field is constant:

$$\frac{B(0)}{B_o} = \left(\frac{2}{\gamma+1}\right)^{2/(\gamma-1)} \quad (20)$$

For the one-dimensional situation, the current that convects downstream will therefore only be a fraction, $B(0)/B_o$, of the total current (which corresponds to the initial field B_o). This fraction is 0.42 for $\gamma=5/3$ and 0.44 for $\gamma=2$. The important implication here is that less than half of the total current will be delivered to a downstream load by a vacuum opening switch if processes upstream of the contact (x=0) conspire to provide a plasma flow at levels equivalent to the semi-infinite reservoir in the present idealization. The observations in some experiments of downstream currents that are about half the total circuit current, and distributions of current density that become broader with convection downstream, thus indicate failure of the switch to open (beyond drawing a high current arc in vacuum). Figure 4 displays the relative magnetic field vs time at various downstream positions for expansion into field-free vacuum of a semi-infinite plasma (with $\gamma=2$) in which the initial magnetic field distribution has a

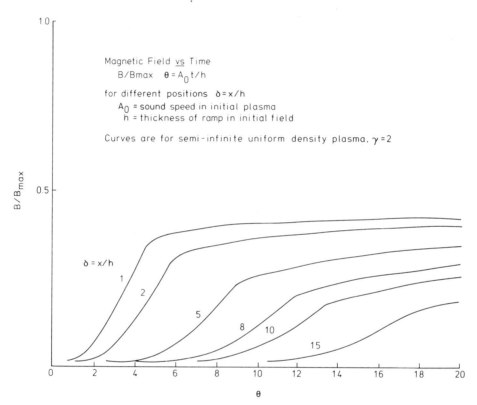

Fig. 4. Suppose the magnetic field in a semi-infinite, uniform density plasma ($-\infty < x < 0$) increases linearly from $x=0$ to $x=-h$, from $B=0$ to a uniform value B_0 (for $-\infty < x < -h$). If this plasma expands adiabatically, with specific heat ratio, γ, into a field-free vacuum, magnetic field will be convected to downstream locations ($x>0$). For $\gamma=2$, the magnetic field is shown as a function of time measured in units of h/a_0, where a_0 is the sound speed in the (uniform field) undisturbed plasma; downstream location is given in units of h. Note that magnetic probe data would indicate that less than half the total current propagates downstream.

linear ramp (-h<x<0) followed by a constant level (-∞<x<-h); distances are normalized by h and times by h/a_o. The curves would be similar to magnetic probe oscillograms when plasma closure, restrike, contact burning or some other plasma source prevents switch opening.

It should be noted that the frozen-field approximation implies that the electromagnetic power flux is limited by the plasma flow. That is, from Eq. (15),

$$\bar{S} = \frac{\bar{u}_\perp B^2}{\mu} . \tag{21}$$

For the semi-infinite plasma, the flow speed from Eq. (17) at x=0 is:

$$u_\perp(0) = 2a_o/(\gamma+1) , \tag{22}$$

so the electric field associated with the Poynting vector flux is a back EMF across the flow:

$$E(0) = 2a_o B(0)/(\gamma+1) \tag{23}$$

$$= \left(\frac{2}{\gamma+1} \right)^{\frac{\gamma+1}{\gamma-1}} a_o B_o , \tag{24}$$

for the uniform initial magnetic field distribution. To achieve higher electric fields and higher power flux levels, it is necessary to reduce the upstream plasma density as much as possible. Acceptance of downstream current fractions on the order of fifty percent means acceptance of electric field values considerably below those needed for rapid energy extraction from an upstream inductive store.

If plasma production upstream of the initial plasma discharge is successfully avoided, then an accelerating current sheet should be obtained. This current sheet will evolve as it moves downstream due to the vacuum boundary conditions on both upstream and downstream sides of the sheet. The downstream side will initially behave like the semi-infinite case. A precursor will expand downstream at a speed given by Eq. 16 for $\rho/\rho_o = 0$.

$$u_1 = \frac{2a_o}{\gamma-1} , \tag{25}$$

carrying with it some magnetic field (i.e., current). Waves traveling upstream through the thickness Δx of the plasma discharge will reflect from the upstream interface between the plasma and a vacuum region in which $B = B_0$. Since the upstream magnetic pressure (in this idealization) is constant, the expansion waves must reflect as compression waves. Within the acoustic formulation, such reflection will give the interface an additional increment of downstream velocity equal to that of the expansion wave. The interface will therefore accelerate through the centered-expansion fan as shown in Fig. 5. The interface represents the position of $B = B_0$, i.e., full current, and overtakes the lower magnetic field distribution that has followed the precursor downstream. The explosion and re-compression of the plasma discharge occurs over a distance scaled by the initial thickness of the plasma, (about fifteen thicknesses in the present simple formulation).

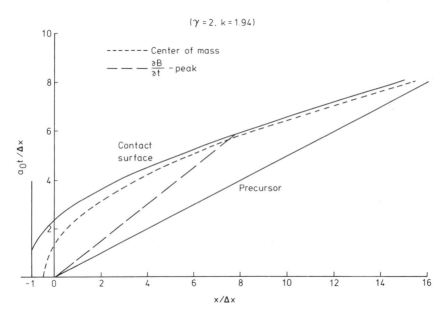

Fig. 5. If the initial plasma density in Fig. 3 is reduced to zero for $x < -\Delta x$, then expansion waves reaching this interface ("contact surface") between vacuum magnetic field and magnetized plasma will reflect as compression waves. Such reflection doubles the flow speed at the interface (compared to the semi-infinite plasma case) so the contact surface accelerates downstream. The arrival of the contact surface corresponds to propagation of the full current downstream. Note that the width of the magnetic field distribution (i.e., the discharge) increases and then decreases as the contact surface overtakes the precursor.

If the load region (or next switch stage) is too close to the switch, then some current may be transferred to the load before the process of re-compression can occur. Degradation of the load may result from the early arrival of a fraction of the total current. Placement of the load region further downstream than necessary, however, can result in lower system efficiency because more kinetic energy is given to the accelerating plasma discharge. Also, more time is available for spurious phenomena to develop. Such phenomena include UV-ablation of surfaces and growth of Rayleigh-Taylor instabilities in the plasma discharge (due to a greater number of wavelengths of travel), resulting in higher upstream densities.

The foregoing remarks and idealized calculations have been presented in a rather broad form to emphasize the common concerns with a variety of plasmadynamic opening switch embodiments. The ability of plasma to convect enough current to affect load and switch performance, while not convecting all the current, was indicated from the frozen-field analysis. Resistive diffusion is important in the low density precursor plasma, so the downstream spread of current and the concomitant loss of action at the main portion of the switch mass can be worse than estimated here. This faster spread and greater loss of action exacerbates the principal difficulty with plasmadynamic switching: the tendency for low density plasma to outrace high density plasma needed during energy accumulation, thereby coupling prematurely to the load (or next switch stage) before the high density slower material has been cleared from the switch region.

THE PLASMA FLOW SWITCH

A successful arrangement for a plasmadynamic opening switch is shown schematically in Fig. 6. Coaxial electrodes are connected in a circuit with an inductive store by means of a radial discharge through a high density plasma annulus. There are several possibilities for creating the high density plasma discharge in vacuum, but experimental work has principally utilized electrical explosion of an array of fine metal wires just upstream of a dielectric foil. The uniform cross-section of the wires helps to achieve simultaneous vaporization and ionization, while the dielectric foil prevents a precursor of wire plasma from spreading downstream before the wire plasma discharge can become established. By judicious choice of wire arrangement and foil thickness, a $(1/r^2)$-mass density distribution can be approximated which simple theory (e.g., magnetoacoustic model) indicates is necessary to avoid mass redistribution. With such a distribution, the motion of the plasma discharge can be calculated using simple zero-dimensional dynamics coupled to a lumped circuit model. A comparison of theory and experiment is shown in Fig. 7. Current vs time (inferred from

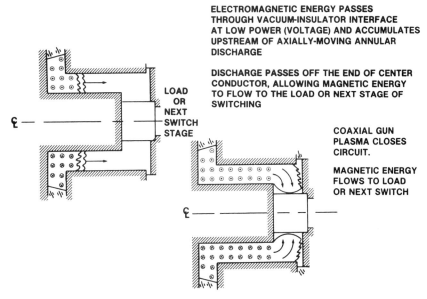

Fig. 6. Schematic of plasma flow switch operation. The inertia of the annular plasma keeps the initial dynamic impedance (rate of change of inductance) low so high currents can be achieved with modest input voltages. Subsequent acceleration of the plasma allows the discharge to sweep off the end of the gun at high speed, permitting sub-microsecond transfer times to the load (or next switch) region.

magnetic probe records) is displayed in Fig. 8 at various stations along the outer conductor of the coaxial plasma flow switch.

The opening switch action for a plasma flow switch occurs when the annular plasma sweeps axially off the end of the center conductor. High mass density plasma continues to move axially at speeds of about 70 km/s. The low mass density plasma upstream of the plasma discharge, however, has sufficiently high sound speeds (given approximately by the Alfven sound speed, $a_0(z) = B(z)/(\rho(z)\mu)^{1/2}$) that the plasma can turn the corner and convect magnetic energy radially inward to an imploding plasma liner. Expansion of low density plasma around the corner of the center conductor converts magnetic energy into kinetic energy of the plasma flow. The power flux is associated with the local electric field that is given by the product of the flow speed and the local magnetic field. Re-compression of the plasma flow as it collects on the boundary of the imploding liner load results in conversion of kinetic energy back into magnetic field energy. The pressure of

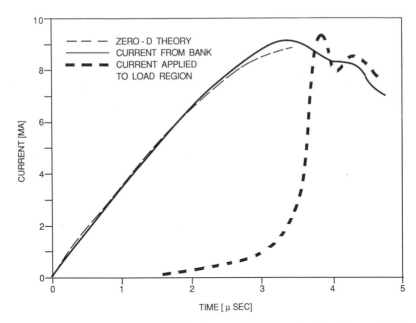

Fig. 7. A zero-dimensional ("slug model") calculation of circuit current versus time (light broken line) compares well with experimental result from 3 MJ (68 kV) test shot of plasma flow switch (on Shiva Star capacitor bank[4]). The current delivered to the downstream load region (heavy broken line) rises in about 200 nanoseconds.

the local magnetic field performs work on the imploding liner, again converting magnetic energy to kinetic energy. (Note that in re-compressing the low density plasma as it contacts the load surface, resistive diffusion may allow magnetic flux to be released directly into the load, providing energy transfer to an electrically-conducting liner beyond simply performing pressure work).

CONCLUDING REMARKS

The basic concept of electromagnetic power flow linked to the motion of high conductivity plasma applies to a variety of switches that create ionized material in vacuum. The particular embodiment of the plasma flow switch accepts as an initial (conservative) design premise that magnetic energy transfer will be limited by plasma convection; its success depends on developing a flow in which the Alfven speed is much greater than the convective speed required to supply flux to the load.

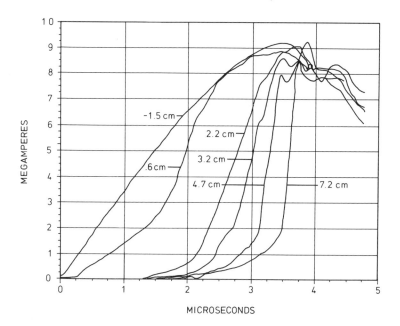

Fig. 8. Current vs time as inferred from magnetic probes stationed along the length of the coaxial gun (in experiment of Fig. 7). Distances are axial displacements from initial discharge position.

As the plasma density decreases (in order to achieve high Alfven speeds), the condition of high electrical conductivity may fail due to insufficient charge-carrier density. The appearance of "anomalous" resistivity will release magnetic flux from the plasma (violating the frozen-field assumption). Power flow to the load will not be limited by convection of plasma if the magnetic Reynolds number is not large. The back EMF across the flowing plasma will be supplemented by the resistive term ηj, where η now has a much higher value than classical estimates, due to particle-wave interactions. Furthermore, as the density decreases, the importance of tensor conductivity (i.e., electron Hall effect) increases. The electric field sustained across a gap will then approach the limit based on magnetic insulation in which the back EMF becomes the electron flow velocity crossed with the magnetic field. Thus, at low plasma densities, the concepts of power flow coupled to single-fluid hydrodynamics, transition to considerations of particle motions and sheath phenomena embodied in the plasma erosion switch.

ACKNOWLEDGEMENTS

The development over the last few years of plasma flow switching at multimegampere current levels has involved the efforts of many people in several organizations including: at R & D Associates, G. Bird, C. Boyer, D. Conte, J. Davis, and S. Seiler; at the Air Force Weapons Laboratory, W. L. Baker, J. Degnan, J. D. Graham, K. E. Hackett, D. J. Hall, J. L. Holmes, E. A. Lopez, D. W. Price, R. Reinovsky, R. Sand, and S. W. Warren.

REFERENCES

1. P.J. Turchi, G. Bird, C. Boyer, D. Conte, J. Davis, L. DeRaad, G. Fisher, L. Johnson, A. Latter, S. Seiler, D. Thomas, W. Tsai, and T. Wilcox, Development of Coaxial Plasma Guns for Power Multiplication at High Energy, Proc. of 3rd IEEE Pulsed Power Conference, 455, IEEE Cat. No. 81CH1662-6 (1981).
2. S.W. Seiler, J.F. Davis, P.J. Turchi, G. Bird, C. Boyer, D. Conte, L. DeRaad, R. Crawford, G. Fisher, A. Latter, W. Tsai, and T. Wilcox, High Current Coaxial Plasma Gun Discharges through Structured Foils, Proc. 4th IEEE Pulsed Power Conference, 346, IEEE Cat. No. 83CH1908-3, (1983).
3. P.J. Turchi, Magnetoacoustic Model for Plasma-Flow Switching, Proc. 4th IEEE Pulsed Power Conference, 342 Cat. No. 83CH1908-3, (1983)..
4. W.L. Baker, G. Bird, J.S. Buff, C. Boyer, C.J. Clouse, S.K. Coffey, D. Conte, D.W. Conley, J.F. Davis, J.H. Degnan, D. Dietz, M.F. Frese, J.D. Graham, S.L. Gonzalez, K.E. Hackett, D.J. Hall, J.L. Holmes, E.A. Lopez, W.F. McCullough, R.E. Peterkin, D.W. Price, R.E. Reinovsky, N.F. Roderick, S.W. Seiler, P.J. Turchi, S.W.R. Warren, and J.M. Welby, Multi-Megampere Plasma Flow Switch Driven Liner Implosion, Megagauss Technology and Pulsed Power Applications, C.M. Fowler, R.S. Caird, and D.J. Erickson, eds., Plenum Press, NY (1987).

FUSE OPENING SWITCHES FOR
PULSE POWER APPLICATIONS

Robert E. Reinovsky

Los Alamos National Laboratory
Group M-6
Los Alamos, NM 87545

INTRODUCTION

   The high power fuse represents a practical opening switch concept that is routinely used in circuits where currents can exceed 25 Megamps[1] and in which the time scales range from 100's of microseconds[2] down to 10's of nanoseconds.[3] The fuse is an electrical conductor which experiences a very rapid rise in resistance as a result of ohmic heating. The heating is driven by the current which the fuse is intended to interrupt, and this heating leads to melting and eventually to vaporization of the conductor. The fuse is, thus, fundamentally a very reproducible one-shot device, and may, within limits, be thought of as totally passive.

Physical Description.

   Figure 1 provides a description of a generic fuse element. The fuse consists of a (relatively thin) conductor which is usually a foil or an array of wires, some form of (relatively heavy) terminal block at each end, and with some medium which may be a bulk solid, a granular solid, a liquid, a gas, or even vacuum, surrounding the conductor. The overall configuration of the ensemble may be flat or cylindrical, or may be tailored to fit special geometries. The fuse interrupts current by virtue of a rise in resistance which is brought about by the heating and ultimate (destructive) vaporization of the thin section of the conductor. Since the temperature rise in the conductor results from ohmic heating, and since that heating is driven by the primary current flowing through the fuse, the details of the performance of a fuse can be a relatively strong function of the circuit in which it is used.

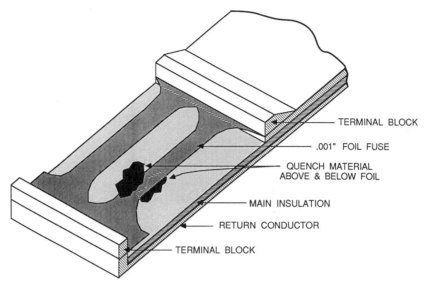

Fig. 1. Conceptual arrangement of generic fuse consisting of fuse conductor, terminal blocks, and quench medium.

The construction of the fuse, specifically the material properties of the conductor, and the geometry of the fuse, are the parameters used to tailor the behavior of the fuse from the beginning of current flow to the time of vaporization. Subsequent to vaporization, the metal vapor interacts with the surrounding medium, and the nature of that medium is important in determining the nature of the interaction. A special case which emphasizes the effect of geometric parameters on the operation of the fuse is a fuse of time varying geometry. Here, one (or both) dimension transverse to the current flow, or even possibly the length parallel to the flow, is varied rapidly (such as by explosive action) thus dramatically changing the resistance/time history of the element.

The family of parameters employed for describing the performance of opening switches are: the voltage supported by the switch during current interruption, the rate of rise of recovery voltage after interruption (RRRV), the maximum allowable conduction time and the minimum achievable interruption time. The voltage supported during interruption is readily described in terms of the time dependent circuit current, the resistivity of the fuse material and its geometry. The voltage can be significantly influenced by the relative timing of the (falling) circuit current and the (rising) fuse resistance. Additionally, the peak voltage is subject to a limitation imposed by the maximum electric field which can be supported by the vaporized conductor. This limit on the voltage scales <u>directly</u> with the length of the fuse. The rate of rise of recovery voltage (RRRV), on the other hand, is determined by the post-vaporization interaction of the fuse material with the surrounding medium.

The maximum time during which the fuse can carry current without a significant rise in resistivity is a function of the equation of state (EOS) of the material, and is, to a lesser extent, a function of the surrounding mediums ability to conduct heat away from, or apply pressure to the conductor. Likewise, the time interval during which the major excursion in resistance occurs, the interruption time, is a function of the material EOS and of the rate at which the circuit can deliver power to heat the material. This is a non-trivial concern when dealing with circuits which switch currents into a low impedance parallel path. The interruption time is a function of the surrounding medium, less through the media's ability to affect the flow of heat to and from the fuse, than through their impact on the expansion dynamics of the conductor material during and after vaporization. It is not surprising that the interrelation of conductor material, circuit current, and the physical properties of the medium leads to a situation characterized by a ratio of conduction to interruption time which changes little for a relatively general class of fuse.

"First Principles" Description of Fuse Operation.

To understand or predict the performance of circuits, such as the standard capacitor/inductive store and switching circuit in Fig. 2, it is necessary to describe the time history of the resistance of the fuse. A convenient, and widely used, approach is to describe the bulk resistivity, ρ, of the fuse material from solid through vapor phases as a function of the state variable, the specific internal energy, e. The resistance is then a simple function of the state of the material and the geometry of the fuse characterized by its length, ℓ, and cross section, A, as

$$R = \frac{\rho(e)\ell}{A} \qquad (1)$$

We immediately observe that, even assuming perfect and complete knowledge of ρ as a function of energy, there is considerable uncertainty in the geometric parameters as a function of time, most notably the dimension of the cross-section during and after the vaporization. Such uncertainty in A(t) makes predicting R(t) based on the circuit and the fuse material a difficult problem -- and as a consequence the process of choosing a desirable combination of fuse and medium parameter becomes even more difficult.

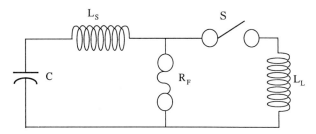

Fig. 2. Capacitor/inductor store circuit typical of many operational fuse switched inductive pulse compression systems. The (left) single loop version is frequently used to develop empirical resistivity information.
(C-storage capacitor, L_S-storage inductor, R_F-fuse
S-closing switch, L_L-load inductor)

These complexities notwithstanding, useful first principles analyses have been published[4,5] for the simplest class of circuits, the RLC circuit in Fig. 2. For this analysis, the resistivity of the fuse material is specified simply as a continuous function of temperature (or internal energy at constant specific heat) defined by a few tabulated values over the range from room temperature up to (but not through) vaporization. Further, we assume that the resistivity is independent of any changes in geometry of the conductor, and that the sinusoidal current waveform is not significantly effected until the onset of vaporization. Then we can integrate the circuit equations using the energy, E, stored in capacitance, C, at voltage, V, the circuit inductance L_S and a constant K_v describing the mass density scaled integral of the resistivity of the material from initial temperature to the onset of vaporization. The appropriate cross-section, A is given by

$$A^2 = \frac{E^{3/2}}{VL_S^{1/2} K_v} . \qquad (2)$$

For fuses of this cross-section, the specific energy accumulated at the time of the peak of the (unperturbed) sinusoidal circuit current is just equal to that required to bring the fuse to vaporization temperature.

The vaporization process requires that an additional amount of energy equivalent to the specific energy of vaporization, ξ, be delivered to the fuse to change the material from a hot liquid (with modest resistivity) to cool vapor (with very high resistivity). For a fuse of length, ℓ, and with liquid mass density, ρ_m, this energy amounts to:

$$E_{vap} = \rho_m \ell A \xi \qquad (3)$$

For experiments aimed at characterizing fuse behavior, the entire stored energy (less that needed to heat the fuse to temperature) is available to drive the material through vaporization, since there is no load in parallel to the fuses. For the more interesting case where energy is to be delivered to a load, as in the circuit in Fig. 2, the amount of energy which must be dissipated in order to achieve flux conservation in the transfer process is known to be $EL_L/(L_S + L_L)$. The fuse is clearly the element best suited to dissipate that energy, and by setting this required dissipation equal to E_{vap} in Eq. (3) we find a criterion on the fuse length:

$$\ell = \frac{EL_L}{\rho_m A \xi (L_S + L_L)} \qquad (4)$$

By substituting Eq. (2) for the cross-sectional area, A, in Eq. (4) we find the length of the fuse is proportional to:

$$\alpha = k_V^{1/2}/\rho_m \xi \qquad (5)$$

The the quantity alpha, which is a function only of intrinsic material parameters, can be taken as a figure-of-merit for fuse materials.

Table 1 summarizes the material properties of various candidate fuse materials, where K_v reflects the energy/resistivity behavior of the material and the values of alpha are seen to vary surprisingly little from one material to another, with silver being the most attractive. In addition to this first principle approach to the choice of materials, more insight can be found by referring to the empirical data which have been accumulated from exploding bridge wire experiments which describe the resistivity as a function of energy for a wide variety of materials.[6] Obviously a material of choice would be one with very low room temperature resistivity and a large change in resistivity for a small increment in energy. Aluminum and copper fall in this category with silver being less attractive despite its low initial resistivity.

TABLE 1

FUSE MATERIAL PROPERTIES

Material	ρ_m	K_v	ξ	α
	kg/m^3	(MKS)	Joules/kg	
Silver	10.5×10^3	3.9×10^{16}	2.3×10^6	8.1×10^{-3}
Copper	8.9×10^3	5.9×10^{16}	4.7×10^6	5.8×10^{-3}
Gold	19.3×10^3	3.0×10^{16}	1.7×10^6	5.2×10^{-3}
Aluminum	2.7×10^3	2.2×10^{16}	10.8×10^6	5.1×10^{-3}

These first principles based observations show:

i) that the cross-section chosen for the fuse typically controls the time of current interruption,

ii) that the length determines the total energy needed to operate the fuse and is to some extent a parameter available to adjust the total resistance (during both heating and vaporization), and

iii) that the nature of the medium determines the hold-off voltage and interruption speed.

This provides sufficient insight to determine the initial parameters for fuse investigations. It is important to note that, except for the bridge wire data which are for cases which do not specifically seek to interrupt current, there are no simple insights into the behavior of the resistivity of materials in fuse-like situations during and after the vaporization process when many more complicated processes occur. This post-vaporization phase is a very important one for practical applications where the fuse is called upon to support voltages and maintain high resistivities over sufficiently long times to prevent restrike.

FUSE MODELING

The value of first principle descriptions of fuse behavior rests primarily on the ability of these models to predict the performance of circuits containing fuse elements with enough fidelity to cover a wide variety of practical applications of fuses.

Guided largely by these first principle approaches, numerous experimental systems over the last 20 years have employed fuses as routinely operating elements in power conditioning systems for beam and plasma physics experiments[7-12] at modest currents and energies. These successful applications have been achieved using little more physical knowledge and understanding than that described in the previous section.

In part, the success of these applications is attributable to the fact that these first principle models do accurately represent most of the fuse's resistance behavior from room temperature through solid heating, melting, liquid heating and into the beginning of vaporization. The model makes no claim of being able to describe the trans- and post-vaporization phases of fuse behavior. While that behavior may be relatively unimportant in low energy systems, it is apparent that vapor phase resistivity dominates the ultimate performance of fuse-containing circuits which are used to deliver large amounts of energy to low impedance loads.

Empirical Models

After the first principle descriptions, resistivity models developed from experimental data have been most widely used for both describing fuse behavior and for projecting circuit performance into extended parameter regimes. These empirical models usually consist of tabular representations of the "effective" material resistivity as a function of either accumulated specific energy, or the related, but not synonymous quantity, "action", and are usually the result of rather careful measurement of the electrical performance of a circuit containing a fuse for which the conditions of physical construction, the details of the surrounding medium, the discharge parameters, and especially the timescale are carefully controlled.[8,13-16]

Measurements of voltage and current in a (usually) single loop circuit provide the raw data from which both resistance and the accumulated internal energy in the fuse, as a function of time, can be determined. Such analysis is conducted subject to the assumptions that the current density distribution, and hence the inductance, in the fuse and in the circuit do not significantly change as a function of time. For convenience, the energy data are usually expressed as "specific internal energy" by scaling the dissipated energy with the fuse mass. The resistance data are expressed as "effective resistivity" by scaling the resistance with geometry, as in Eq. (1), while making the superficial assumption that the cross-section of the material does not change appreciably over times of interest.

Figure 3 shows representative voltage and current data from one of the many experiments[12] reported in the literature. The fuse

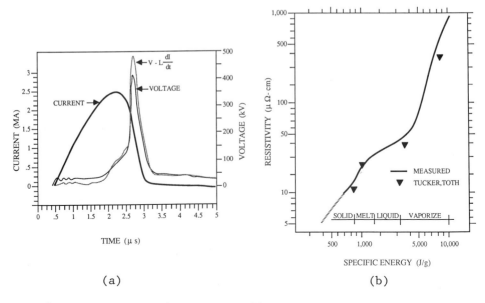

Fig. 3. Current, voltage and "effective resistivity" for aluminum fuses.[12]

was made of aluminum and the surrounding medium was small diameter glass beads. Figure 3a shows the current measured in the fuse, the voltage measured across the physical terminals of the fuse, and the voltage after the inductive component of fuse voltage (L dI/dt, with L independent of time) has been subtracted. The resistance is found from the ratio of V/I, and when scaled by the length and initial cross-sectional area, becomes the "effective resistivity" shown in Fig. 3b. Similar data have been reported for different fuse and quenching materials and some of these data are discussed in the next section. As expressed, the effective resistivity should, of course, be a "state variable", completely describing the conductor material, and implicitly the interaction of the fuse material with the surrounding quenching material. In general, the literature contains data showing "effective resistivity" behaviors that are comparable for comparable physical situations to within 10% during the solid and liquid phases and which differ by no more than factors of several during the vaporization phase.

Such tabular data are readily incorporated in circuit analysis studies of fuse-based inductive pulse compression systems. The energy dissipation in the fuse is calculated in a finite difference solution to the circuit equations, and the resistivity associated with that energy is then scaled and used to determine the resistance of the fuse for the next iterative step. The principal advantage of the empirical data for circuit analysis studies is that it describes the behavior of the fuse and hence of the circuit into and through the vaporization phase and provides a tool which allows predictions of circuit performance, particularly when driving dynamic loads.

Despite the usefulness of the empirical data as a predictor of circuit behavior, uncertainty in performance during vaporization is a severe handicap when attempting to find ways to significantly improve the performance of fuses.

Heat Transport Models

By adding treatment of one dimensional heat flow to the first principles description and to the empirical models of fuse operation, an extended model which begins to account for some of the effects related to the surrounding (quench) material emerges.[17] As in the first principles description, the heat transport model makes no attempt to describe the motion of material before or during vaporization but unlike the basic description, both heat transport by conduction into the quenching material, and 1-D non-uniform, time dependent temperature profiles within the conductor can be considered. As in the simple model, resistivity is taken to be a function of temperature, but unlike the simple model, the heat capacity used to relate internal energy to temperature can also be a function of temperature. We have noted that the comparison between the empirical model and "handbook" resistivities is quite good in solid and liquid phases, and therefore the empirical model may be reasonably extrapolated to temperatures well below those explored in the experiment by using handbook data. One such case of particular interest is aluminum in the range of liquid nitrogen temperatures.

The 1-D heat transport model was developed to explore the impact on overall circuit performance that could be expected from:

i) cooling the fuse conductor and surrounding quench to cryogenic temperatures or,

ii) introducing quench materials which have very high values of thermal diffusivity, and hence could be expected to remove heat from the fuse rapidly.

Figure 4 shows a comparison of the calculated current and voltage behavior for aluminum fuses operated in a conventional manner; for a fuse cooled to liquid nitrogen temperature ($70°$ K), and a cooled fuse operated in a medium with highly enhanced thermal diffusivity (the square root of the product of mass density, heat capacity and thermal conductivity). For this calculation, empirical resistivity data, such as in Fig. 3, were used when the fuse temperature was above the melting temperature, and a finite differ-

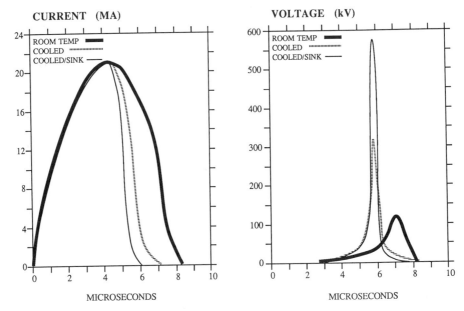

Fig. 4. Calculated performance of aluminum foil fuses operated under nominal conditions ("Room Temp"), when cooled to liquid nitrogen temperature with quartz quench, and when cooled with SiC quench.[17]

ence, thermal diffusion model was used below room temperature. In each case, the fuse length and cross-section were chosen to allow the circuit to deliver the same (20 MA) current to the storage inductor. The plots show the advantage gained by cooling and heat sinking the fuse; namely more rapid interruption of current and correspondingly higher voltages. These improvements result primarily from the reduced cross-sectional area, and increased length that are possible with reduced initial resistance and enhanced heat flow. Little improvement was found for the case of enhanced diffusivity quenching material at room temperature, however.

Magneto-Hydro-Dynamic (MHD) Models

Both one and two dimensional MHD models have been employed to describe the performance of fuses. MHD models have the advantage of potentially adding information concerning the dimensional behavior of the fuse material during the vaporization phase when dynamic effects and changes in density and temperature play a significant role in determining the state of the material.

One dimensional models linked to elementary equation of state (EOS) data have been employed to describe the motion of the fuse material up to and through vaporization.[18] Such models provide reasonable descriptions of fuse configurations involving foil-like fuses and solid quenching materials, and can be readily applied to gas and vacuum geometries as well. Some insight drawn from these calculations can also be applied to more complicated (and more widely used) geometries and materials. Calculations employing EOS data with realistic values for the compressibility of insulator materials, such as those presented in Fig. 5, reveal significant motion of vapor/insulator interfaces (up to several tenths of a millimeter) under the combined influence of the material (hydrodynamic) pressure from the exploding conductor and the magnetic pressure gradient resulting from the unavoidably asymmetric magnetic field which pushes the conducting channel before it. The combined effects of reduced pressure and density and increased effective cross section, portrayed by the zone-position vs. time plot in Fig. 5a, can be interpreted as an "effective resistivity", as shown in Fig. 5b, thus providing some insight into the processes which limit the performance of fuses when confined by solid insulators. Thus the figure indicates the extent to which the simple-minded "effective resistivity" models are affected by hydrodynamic motion.

While such models are important in gaining understanding of the fusing process, their quantitative usefulness is severely limited by the validity of the EOS data which they access, and by the fact that they are incapable of addressing the mixing of fuse material with surrounding porous media -- the approach which continues to be the most promising one experimentally.

Two dimensional models have also been employed, but to date their use has been limited to exploring single exploding wire configurations where "hot spots", MHD instabilities, and anomalous resistivity effects dominate. While two dimensional calculations of foil fuses in structured media are possible, they are obviously very time consuming and without significantly improved EOS data it is unlikely that they will add significantly to the fundamental level of understanding of the subject and are therefore not generally being pursued.

Advanced Equation of State Models

Perhaps the most significant result from modeling efforts in the last several years is the application of advanced equation of state data to fuse situations. One such recent effort[19] has compiled equation of state information from several sources and produced a model which more fully characterizes the low temperature conductivity of aluminum metal vapor. The consolidated model has

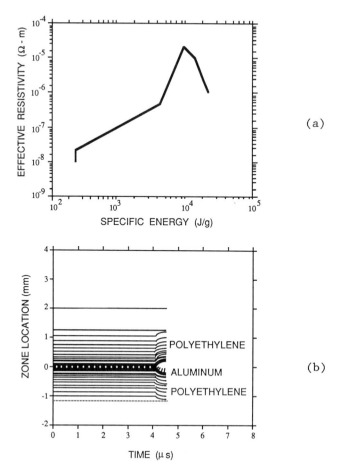

Fig. 5. One dimensional MHD calculation of foil fuse in semi-infinite polythylene quench. The zone position plots (from the Lagrangian code) show the formation of a void which fills with low density vapor. The top plot shows voltage and current information from the code interpreted as "effective resistivity"[18].

been linked to a one-dimensional MHD calculation and the results have been compared with experimental measurements. A useful presentation of such a model is shown in Fig. 6 where resistivity contours (normalized to initial resistivity) are plotted in a density/temperature parameter space. The trajectory calculated for a thin aluminum fuse through density/temperature space is shown on the plot. The figure enables some understanding of the important role that material properties play in controlling fuse performance.

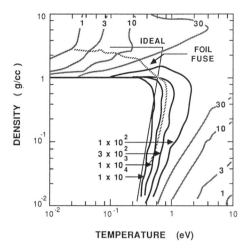

Fig. 6. Resistivity contours normalized to initial resistivity in density/temperature space for aluminum, showing trajectory followed by an exploding foil (not interrupting current) and an "ideal" fuse.[19]

While even the improved EOS data do not claim to fully characterize the solid, melt, and liquid regions accurately, inspection of the density-temperature space give, for the first time, some quantitative insight into an approach to improve the performance of the fuse after entering the vapor phase. As indicated by the "ideal" trajectory, the fuse needs to be thoroughly heated to vaporization temperature while maintaining as high a density as possible, then allowed to expand in a way which drops the density while not allowing the temperature to rise significantly, and in fact cooling of the vapor is clearly helpful. Furthermore, the plot suggests the possibility that resistivities, almost an order of magnitude higher than the "effective resistivities" already observed in empirical data, are possible.

Thus MHD models of such one dimensional configurations, coupled with these improved EOS data sets, suggest that rather significant improvement in performance can be expected if 1-D behavior can actually be achieved. In principle, 1-D expansion of the vapor can be approached if current densities, material properties, magnetic field configurations, and construction techniques are well controlled. The improvement in performance predicted by the MHD codes justifies continued effort in this direction.

FUSE PERFORMANCE

Fuses for pulse power applications, have been explored rather superficially at the basic physics level, more thoroughly surveyed through empirical investigations, and are applied routinely as components of operating systems perhaps more frequently, and with more success, than the current level of basic understanding might justify. High speed fuses have been operated over a broad range of currents from a fraction of a kiloampere to tens of megamperes, and over timescales that range from tens of nanoseconds to milliseconds.

As already noted above, the behavior of a fuse up to the time of vaporization is determined by the geometry and by the material properties of the conductor. There is little opportunity within the limited flexibility offered by these parameters to affect the overall performance of the fuse significantly. After vaporization, however, the metal vapor interacts with the surrounding medium. The extent to which that medium interrupts the continuity of the vapor channel or cools the channel by virtue of its heat capacity or thermal conductivity; the extent to which it encourages or inhibits expansion of the vapor and hence alters the density and temperature of the vapor; or the extent to which it adds electrons (by ionization) or removes them (by attachment) can significantly affect the resistivity and the cross-sectional dimensions of the vapor channel and the maximum electric field that the channel can support. Thus we look to the surrounding medium, the "quench", and to the interaction of the vapor with that medium as our primary source of flexibility as we seek to not only understand fuse operation, but to extend fuse performance.

The literature contains reports of the results of numerous, fuse-like, exploded conductor experiments using a wide variety of conductor and quenching materials. Many of the experiments consist of a capacitor discharged through the series connection of an inductor and a fuse. The primary diagnostics are electrical measurements of circuit voltages and current. The analysis of these data is aimed at evaluating the resistance implied by these quantities. Thus, in characterizing the experiments, a charging time (the quarter period of the circuit), a peak voltage across the fuse (usually not corrected for inductive voltage) normalized to the capacitor charge voltage, a peak fuse resistance normalized to the fuse's initial resistance, a maximum electric field across the fuse (peak voltage/length), and a time compression factor are reasonable parameters for comparing the performance of various configurations.

Table 2 attempts to summarize a variety of fuse experiments and their results. For most of these cases the data pertains to fuses which are fed with continuous current during their operation (i.e., no current is switched into a parallel circuit) so that the heating and vaporization process is driven to completion. This configuration avoids the risk of underdriving the fuse, and achieving incomplete vaporization. Futhermore, in most cases, sufficient energy is available to heat the vapor to a temperature where significant thermal ionization takes place, leading to voltage breakdown and thus clearly defining one limit of performance. It is important to note that the community has explored a wide variety of techniques at a variety of energy levels, and that the result in Table 2 are intended to be representative and not to report the absolute limits for various techniques.

Fuses in Solid Media

Fuses operated with solid quenching materials may be divided into two categories. The most frequently used configuration is that of a foil fuse in a granular material. The grain size and composition of the quench may vary significantly, and the primary interaction mechanism is that of the relatively freely expanding vapor mixing with the grains of quenching material. The alternate approach is a solid quenching material in contact with, or spaced slightly away from the conductor, in which the primary interaction is that of confinement of the metal vapor by interaction with the more or less solid and immovable walls. A variant of the latter approach is the use of solid quenching material in the form of a thin sheet of insulator which is driven away from the fuse by the hydrodynamic pressure of the vaporizing conductor and whose mass is chosen to control the rate and magnitude of vapor expansion in a "tamper-like" interaction.[19]

Among materials used as granular quenches, conventional (soda-lime) glass beads and quartz beads and powders with grain sizes ranging from sub-microns to many hundreds of microns diameter are the most common ones. In addition some data have been gathered using commercially available alumina (Al_2O_3) and silicon carbide. Glass beads, commercially available as material for "sand-blasting" operations, have the advantage of economy, remarkable uniformity, and ready availability. They have the possible disadvantages of relatively low melting temperature. Furthermore, they contain appreciable quantities of sodium and potassium, both of which are alkali metals with low first ionization potentials. The alkali metals are readily ionized at metal vapor temperatures, and electrons resulting from this thermal ionization may contribute to lowering the fuse's final resistivity. Crushed quartz or quartz beads, on the other hand, are somewhat more expensive and are available from relatively few sources but have the off-setting advantages of higher melting temperature and remarkably high purity, containing only silicon and oxygen (SiO_2) in an insulator of high integrity. In the US, systems, even those in the 5-10 MJ class which are operated routinely, have tended to employ glass bead quenches. In the Soviet Union, experiments tend to employ smaller grain size quartz powder but much of the Soviet data are for the 100 kA to 1 MA, few microsecond, regime.

TABLE 2

SURVEY OF FUSE PERFORMANCE

	I_p (MA)	V_p/V_0	T_{ch} (µs)	T_{in} (µs)
FOILS IN GRANULES				
100 µ Glass	30	4	5.0	0.30
1 µ Quartz	0.4	4	3.5	0.10
Glass	0.3	4	4.0	0.50
Alumina	0.3	4	4.0	0.50
Silicon Carbide	0.3	4	4.0	0.50
FOILS IN SOLID				
Polyethlene	0.4	1.5	3.0	0.45
Paraffin	0.4	1.6	3.8	0.60
Mylar	0.4	5.0	4.0	0.40
FOILS IN LIQUID				
Water	0.4	2.0	3.7	0.70
Water	0.3	4.0	4.0	0.50
Water	0.008	23.	200.	8.
Water(cylind)	0.1	10.	200.	4.
WIRES IN LIQUID				
Water	0.005	2.4	2.5	0.25
Water(tube)	0.0005	70.	1000.	–
Water	0.1	–	1.7	–
FOILS IN GAS				
Air	0.4	2.6	3.2	0.20
WIRES IN GAS				
N_2 @ 10 Atm	0.05	3.0	2.8	0.10
Air @ 1 Atm	0.01	–	1.0	0.20
WIRES IN VACUUM				
1×10^{-5} Torr	0.01	–	1.0	0.20

T_{ch}/T_{in}	R_p/R_o	E_m (kV/cm)	Reference
17.	350	5.0	Reinovsky(1)
35.	200	5.4	Burtsev(20,21)
8.	200	4.4	Bueck(15)
8.	225	4.5	Bueck(15)
8.	100	4.2	Bueck(15)
6.6	100	1.6	Burtsev(20,21)
6.3	170	1.8	Burtsev(20,21)
10.	160	4.0	Goforth(22)
5.3	320	2.0	Burtsev(20,21)
8.0	230	4.3	Bueck(15)
25.	150	3.3	Conte(9)
5.0	-	8.2	Wilkinson(24)
10.	-	14.	Salge(23)
-	-	1.5	Salge(23)
-	-	13.5	Conte(2)
16.	80	2.7	Burtsev(20,21)
28.	-	7.5	Kotov(3,11)
5.	-	20.	Vitkovitsky(26)
5.	-	9.0	Vitkovitsky(26)

Both alumina and silicon carbide are also commercially available, as abrasives, in granular form and in appropriate sizes, and are relatively economical. As discussed in the Modeling section, they are of interest because of their relatively high thermal diffusivity which leads to a high rate of heat flux into the material. Though the previous discussion was directed toward cooling the fuse in the early stages of operation by enhanced heat flow, this property also suggests that such quenching materials may also be good for cooling high temperature vapor. Silicon carbide tops the list of readily available materials, ranked by thermal diffusivity, but its utility is limited because it demonstrates some degree of semi-conductor like behavior at elevated temperature. Experiments using SiC material have demonstrated notably lower resistance multiplication when compared to experiments in which quartz and alumina media are employed. These experiments are illustrated by the data in Table 2.

A relatively large number of experimental results have been reported in which fuses operated in granular quenching materials have successfully interrupted currents. Figure 7a shows one of those results. As a general comment, the highest energy results to date[1,20,21] have been obtained using foil fuses in granular quenching materials. The AFWL Shiva Star data (Fig. 7a) in which fuses, operated at several megajoule energies employing granular glass bead quenching materials, carried current for about 5 microsecond, and interrupted 15-30 MA peak currents against 200-300 kV voltages in 200-300 ns is the highest energy operating system reported in the literature. Considerable empirical data exist for currents up to a few megamps, delivered over times of a few microseconds to a few hundred microseconds, producing inductive voltages upon opening of 100-200 kV. In general, a remarkable similarity exists among all the granular quenched fuse data. Considering the table entries which refer to granular quenched cases, the common aspects of their behavior seems to be a ratio of conduction to interruption time of 15-30, a resistivity multiplication of 200-500 and a peak electric field of about 5 kV/cm produced during the interruption.

Relatively little data have been reported describing the behavior of fuses when confined by solid (sheet or tube) material, and what data exists are not particularly encouraging. This result is consistent with the 1D-MHD modeling results presented in Fig. 5, where expansion of the conductor vapor and compression of the insulator leads to a broad conducting channel resulting in low resistance even if the channel resistivity remains high! However, the more recent modeling work of Lindemuth[19] and experiments by Goforth,[22] suggests that a tamper-like configuration may be an attractive way to control precisely the trajectory of fuse operation through the density/temperature space of Fig. 6, and this is an encouraging possibility.

FUSE OPENING SWITCHES FOR PULSE POWER APPLICATIONS 227

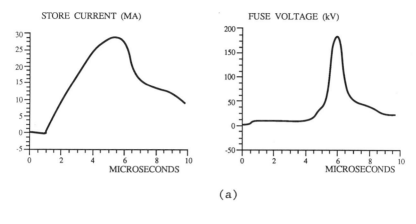

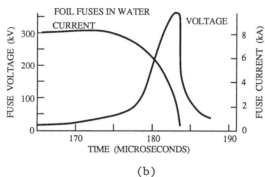

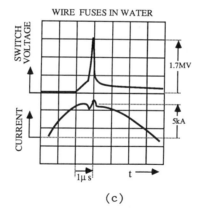

Fig. 7. Operation fuse data.
(a) Shiva Star operated at 6 MJ, and switched into a resistive load,[1]
(b) First state of the Trident experiment,[9]
(c) Low energy wires in water.[23]

Fuses in Liquid Media

Where granular quenching materials offer the possible advantage of enhanced mixing of fuse vapor and insulating media, liquid quenches offer the possibility of achieving uniform, inertial, confinement of the vapor at very high pressure. This is especially true for cases of fuses consisting of an array of small diameter wires where, initially, the rate of volumetric expansion (for a given sound speed in the medium) is much less for a circular geometry than for an equivalent conductor area of flat cross-section. For relatively long time scales (100's of microsecond) data have been reported[2] for foil fuses in water quenches in which the ratio of conduction time to interruption time exceed 20, which is much greater than the factor of 10 usually observed in granular quenched fuse experiments. Other water quenched experiments,[9,23] in which the conductor is a wire array, have shown significantly higher induced electric fields (up to 15 kV/cm) than those demonstrated by the granular quenches. Specially configured water quenched wire fuses have been operated for millisecond times at electric fields up to 1.5 kV/cm. Resistance multiplication, however, were in those cases no greater than those reported for granular quenched fuses.

For comparison with the very high current foil fuses operated in granular quenches, shown in Fig. 7a, the data in Fig. 7b is for an aluminum foil fuse operated in water for a long charging time (>150 μs) which dramatically exhibits the factor of 20 ratio of conduction to interruption time. On the other hand, Fig. 7c shows other low current data taken with a single, long, water quenched wire charged for only microsecond times but demonstrating electric field strengths of 14 kV/cm.

Fuses in Gaseous Media

A large body of literature exists for gaseous quenching media, reflecting the performance of exploding wires in air, various gases and vacuum and a limited sub-set of these data are reported for fuse-like parameters.[25-27] These experiments usually involve a circuit in which more energy is available than that necessary to vaporize the wire, and hence the metal vapor is not only heated, but raised to a temperature where significant thermal ionization occurs -- and breakdown follows.

Several investigations have been conducted comparing the electric field produced during heating and vaporization of a wire in gas and in vacuum, and more importantly exploring the evolution of dielectric strength after the initial current interruption -- experiments which frequently require the reapplication of voltage by an external source. In terms of speed of current interruption, wires in gas and vacuum are quite attractive, but their main drawback is the relatively short duration of their high resistance phase. The role of low density vapor at the leading edge of the expansion front in providing a path for avalanche breakdown and the role of MHD instabilities which disrupt and segment the wire/plasma column have been investigated computationally and theoretically. Both mechanisms appear to contribute to the collapse of electric field strength at early times. The results of some of these investigations are tabulated in Table 2.

Among the interesting implications of the data is the observation that for a relatively low energy array of fine wires there are data showing very short interruption times on the order of 10's of ns. Additionally, some data are reported showing a notably high induced electric field, up to 20 kV/cm, but these results are only achieved in very fast systems with relatively modest time compression. Successful application of high pressure gas quenched fuses have been reported,[3] and such fuses appear to be particularly well suited to very fast, but low current applications such as electron accelerators, where the high but short lived electric field strength of the wire allows practical fuse lengths to be employed.

Rather thorough work has also been reported showing the time evolution of the dielectric strength of wires exploded in gas or vacuum as a function of delay time after the vaporization of material.[26] Additionally, a little similarly quantative work has been reported for foil fuses in a water quench medium.[2]

Dynamic Quenches

Most of the quenching materials discussed are passive in nature, deriving their effects from their ability to absorb heat, to mix with the fuse vapor, or to move in response to the pressure of the exploding foil. However, concepts which include quenching materials which actively participated in the energy budget of the system can be envisioned, and some have been explored and reported.

A variant of the concept of liquid quenching materials is that of introducing a chemically active medium which reacts with the foil conductor and adds energy to the system, and thus presumably hastens the solid to vapor transition. An example of such a system is an aluminum fuse in a relatively high concentration of hydrogen peroxide which has shown some modest impact on the performance of the fuse -- though probably not enough to justify the additional complexity and hazards associated with such techniques.[16] While the concept of a chemically reactive quenching is in principal equally appropriate to solid and gaseous quenching systems, the majority of practical chemical systems are in liquid form.

Another conceptual variant of solid quenching is that of using a plastic bonded high explosive as the quenching material, in which case the HE, if carefully and uniformly detonated over a large area can apply considerable compressive force and maintain the vapor density at a high level. Relatively large changes in the performance of small scale fuses have been both calculated[18] and reported experimentally[28] in a case where the total high explosive energy is orders of magnitude larger than the discharge energy. These concepts are poorly understood at this time.

SUMMARY

This review has attempted to provide some overview into the magnitude and variety of both the work that has been reported exploring the behavior of fast fuse opening switches, and into the variety of configurations and applications in which fuses can be routinely employed. We began by describing the physical construction of a typical fuse element, and then discussed the first principles description of fuse operations. We then considered a variety of modeling approaches by which we can gain some understanding of the processes that limit fuse performance -- of which an empirical model developed from data gathered in small scale experiments are still one of the mainstays. We then described results reported for a few of the experiments reported in the literature which also give some sense of the processes which limit fuse performance.

Fuses appear to have wide applicability in one-shot systems ranging from kJ to 10's of MJ of stored energy. Future theoretical efforts should concentrate on extending EOS information and MHD modeling of the vapor phase behavior and experimental efforts should center on the implementation and measurement of fuses in truly one dimensional configurations.

REFERENCES

1. R.E. Reinovsky, W.L. Baker, Y.G. Chen, J. Holmes, and E.A. Lopez, Shiva Star Inductive Pulse Compression System The 4th IEEE International Conference on Pulsed Power, IEEE Cat. No. 83CH1908-3, NY (1983).
2. D. Conte, R.D. Ford, W.H. Lupton, J.D. Shipman, P. Turchi, and I.M. Vitkovitsky, Inductive Charging of Pulse Lines in 0.1 to 1.0 MJ Range, Using Foil Fuses Staged with Explosively Actuated Switches, NRL Report 3742 (1978).
3. Y.A. Kotov, B.M. Koval'chuk, N.G. Kolganov, G.A. Mesyats, V.S. Sedoi, and A.L. Ipatov, High Current Nano-Second Pulse Electron Accelerator with Inductive Shaping Element, Sov Tech Phys Lett, 3:9 (1977).

4. Ch. Maisonnier, J.G. Linhart, and C. Gourlan, Rapid Transfer of Magnetic Energy by Means of Exploding Foils, Rev Sci Instrm, 37:10 (1966).
5. J.N. DiMarco, and L.C. Burkhardt, Characteristics of a Magnetic Energy Storage System Using Exploding Foils, J. Appl. Phys., 41:9 (1970).
6. T.J. Tucker, and R.P. Toth, A Computer Code for Predicting the Behavior of Electrical Circuits Containing Exploding Wire Elements, Sandia National Lab Report 75-0041 (1975).
7. L.C. Burkhardt, R. Dike, J.N. DiMarco, J.A. Phillips, R. Haarman, and A.E. Schofield, The Magnetic Energy Storage System Used in ZT-1, Proceedings of Int'l Conference on Energy Storage, Compression and Switching (1974).
8. V.A. Burtsev, A. B. Berezin, A.P. Zhukov, V.A. Kubasov, B.V. Lyvublin, V.N. Litunovskij, V.A. Ovsyannikov, A.N. Popytaev, A.G. Smirnov, V.G. Smirnov, V.P. Fedyakov, and V.P. Fedyakova, Heating of a Dense Theta Pinch with Strong Fast-Rising Magnetic Fields, Nuclear Fusion, 17:5 (1977).
9. D. Conte, R.D. Ford, W.H. Lupton, and I.M., Vitkovitsky, Trident - a Megavolt Pulse Generator Using Inductive Energy Storage, Proc. 2nd IEEE Int. Pulsed Power Conference, IEEE Cat. No. 79-90330, NY (1979).
10. H.C. Early, and F.J. Martin, Methods of Producing a Fast Current Rise from Energy Storage Capacitors, Rev Sci Instrum, 36:7 (1965).
11. Y.A. Kotov, N.G. Kolganov, V.S. Sedoi, B.M. Koval'chuk, and G.A. Mesyats, Nanosecond Pulse Generators with Inductive Storage, 1st IEEE Int'l Pulsed Power Conf., IEEE Cat. No. 76CH1197-8 (1976).
12. R.E. Reinovsky, D.L.Smith, W.L. Baker, J.H. Degnan, R.P. Henderson, B.J. Kohn, D.A. Kloc, and N.F. Roderick, Inductive Store Pulse Compression System for Driving High Speed Plasma Implosions, IEEE Trans Plasma Sci, PS-10:3 (1982).
13. A.P. Baikov, L.S. Gerasimov, and A.M. Iskol'dskii, Electrical Conductivity of an Aluminum Foil in an Electrical Explosion, Sov Phys - Tech Phys, 20:1 (1975).
14. J.C. Bueck, and R.E. Reinovsky, The Effects of Cryogenic Initial Temperature on Aluminum and Copper Electrically Exploded Foil Fuses, 4th IEEE Pulsed Power Conference, IEEE Cat. No. 83CH1908-3, 1983.
15. J.C. Bueck, and R.E. Reinovsky, High Performance Foil Opening Switches, 5th IEEE Pulsed Power Conference, IEEE Cat. No. 85C2121-2 (1985).
16. D. Conte, M. Friedman, and M. Ury, A Method for Enhancing Exploding Aluminum Foil Fuses for Inductive Storage Switching, 1st IEEE Int'l Pulsed Power Conference, IEEE Cat. No. 76CH1197-8 (1976).

17. W.F. McCullough, Private Communication (1983).
18. W.F. McCullough, One Dimensional Magnetohydrodynamic Simulations of Exploding Foil Opening Switches, 5th IEEE Pulsed Power Conference, IEEE Cat. No. 85C2121-2, (1985).
19. I.R. Lindemuth, J.H. Brownell, A.E. Greene, G.H. Nickel, T.A. Oliphant, and D.L. Weiss, A Computational Model of Exploding Metallic Fuses for Multimegajoule Switching, J. Appl. Phys., 57:9 (1985).
20. V.A. Burtsev, V.N. Litunovskii, and V.F. Prokopenko, Electrical Explosion of Foils, I, Sov Phys - Tech Phys, 22:8 (1977).
21. V.A. Burtsev, V.N. Litunovskii, and V.F. Prokopenko, Electrical Explosion of Foils, II, Sov Phys - Tech Sci, 22:8 (1977).
22. J.H. Goforth, I.R. Lindemuth, H. Oona, R.E. Reinovsky, K.E. Hackett, W.F. McCullough, and E.A. Lopez, Exploding Metallic Fuse Physics Experiments, IEEE Int'l Conf on Plasma Science, (1986).
23. J. Salge, U. Braunsberger, and U. Schwarz, Circuit Breaking by Exploding Wires in Magnetic Energy Storage Systems, Proc. Int'l Conference on Energy Storage, Compression and Switching, (1974).
24. G.M. Wilkinson, and A.R. Miller, Generation of Sub-microsecond Current Risetimes into Inductive Loads Using Fuses as Switching Elements, 5th IEEE Pulsed Power Conference, IEEE Cat. No. 85C2121-2 (1985).
25. E.V. Krivitskii, and V.P. Litvinenko, Exploding Wire Mechanisms, Sov Phys - Tech Phys., 21:1218 (1976).
26. I.M. Vitkovitsky and V.E. Scherrer, Recovery Characteristics of Exploding Wire Fuses in Air and Vacuum, J. Appl. Phys., 52:4 (1981).
27. G.P. Grazunov, V.P. Kantsedal, and R.V. Mitin, Use of Electric Explosion of Wires in a High Pressure Gas to Break a Current Circuit, Zhurnal Prikladnoi Mekhanikii Tekhnicheskoi Fiziki, 6:102 (1976).
28. P.S. Levi, Explosively Driven Rupture Conductor Opening Switches, 4th IEEE Pulsed Power Conference, IEEE Cat. No. 83CH1908-C (1983).

EXPLOSIVELY-DRIVEN OPENING SWITCHES

B.N. Turman and T.J. Tucker

Sandia National Laboratories
Albuquerque, N.M. 87185

INTRODUCTION

Explosively-driven opening switches may be viewed as evolutionary developments from mechanical circuit breakers, fuses, or gas-quenched arc devices. For example, one explosive switch concept which can be viewed as a development from a mechanical circuit breaker uses an explosive charge to cut a metal conductor with a dielectric knife blade.[1,2] With a solid conductor as the switching element, this switch has the advantage of conducting a very high current density for an unlimited time, until it is command-fired. The hold-off voltage of the switch is principally limited by the time-dependent breakdown strength of the explosive products. The opening time of the switch is dominated by the mechanical movement of the conductor and is limited by the conductor velocity. Ford and Vitkovitsky[1] have reported opening times of about 10 μs and peak hold-off voltage of 35 kV.

A concept which has proven to give faster opening times replaces the solid conductor with a conducting plasma channel. A schematic of this device, first proposed by Pavlovskii,[3] is shown in Fig. 1. Pavlovskii used this design to switch the output current from an explosive generator into a 38 nH inductive load. The peak generator current was 7 MA, and the peak load current was 4 MA with a rise time of 0.5 μs. The active switching element of this design was a cylindrical plasma channel formed from an electrically exploded thin aluminum film (item 2 in Fig. 1). The plasma resistance was about 1 mΩ during operation of the explosive generator current source. The opening switch was activated by detonating the explosive charge (item 3) and propagating a detonation wave outward from the center of the cylindrical explosive. Initiation of the explosive was timed so that the detonation wave reached the outer surface of the explosive and began to compress the surrounding plasma channel at the time of peak current from the generator. The switch resistance rose from 1 mΩ to 200 mΩ in about 0.5 μs. The peak switching voltage was 300 kV, corresponding to an electric field of 25 kV/cm. Metal grading rings on the insulating outer cylinder (item 1 in Fig. 1) graded the electric field across the wall to reduce the chance of high voltage breakdown. Pavlovskii estimated that 100 kV/cm was the breakdown level for the explosive products. As the switch resistance neared peak, a closing switch (item 5) was activated, and the current in the primary side of the circuit was diverted to the load. The risetime of current into the 38 nH load (0.1 to 0.8 of peak value) was 0.45 μs and the peak load current was 4 MA.

Similar switches have been tested by the authors[4] with comparable results. This cylindrical geometry has been used for high current, high energy experiments, and planar geometry tests have been performed for physics experiments and scaling studies.[5] Several modeling efforts have been undertaken at Sandia National Laboratories, Los Alamos National Laboratory and the Air Force Weapons Laboratory to explore the potential and limitations of this concept. A review of these research efforts on the Pavlovskii fast opening switch is the topic of the remainder of this section.

ELECTRICAL CHARACTERISTICS OF EXPLOSIVES

In this switching concept, the explosive gas products are intimately associated with high voltage elements; consequently, the high-voltage stand-off characteristics of the gas products are of importance to any assessment of the performance of such a switch. Additionally, the explosive gas products form a shunt resistance path across the opening plasma switch, so the resistivity of the

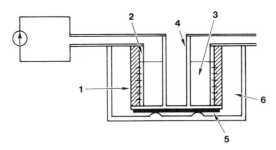

Fig. 1. The "Pavlovskii switch": (1) insulator; (2) aluminum-coated film; (3) switch explosive charge; (4) inner tube of storage unit; (5) insulation of closing device; and (6) load.

explosive products must also be known. Considerable experimental work in this area has been reported, particularly in the Soviet literature.

First, a short description of some basic explosive considerations is appropriate. The detonation velocity of typical explosives[6] is in the range of 3-8 mm/μs. This is the velocity at which the detonation front moves through the explosive gas at the free surface of the detonated explosive. Another parameter of interest is the detonation pressure. Typical values range from some 200-300 kbar. A listing of explosive parameters is given in Table 1 for some of the more common explosives used in the United States. Also listed, where possible, is the equivalent, or comparable, explosive used in the Soviet Union.

A considerable body of data is available from the Soviet literature concerning the electrical resistivity of explosive products. A summary of these data is tabulated in Table 1, as well. In general, the detonation front, where the explosive is chemically reactive and the temperature is high, is observed to be highly conductive. Behind the detonation front, the resistivity rises abruptly, and within 0.5 μs, or some 4 mm behind the front, the resistivity reaches an equilibrium value in the range of 0.2-10 Ω-cm.[7] At late times, on the order of 2 μs after detonation, the resistivity of PETN and RDX are reported to increase to values around 1×10^4 Ω-cm.[8]

The conductivity of detonating explosives is a complex, time-dependent function involving not only the type of material but also the physical state of the explosive used. It appears that powdered explosives are, in general, more resistive at equilibrium than are solid, cast explosives, possibly because of the reduced pressure and density at equilibrium. (Pressure and density within the detonation wave were shown to be directly related to the conductivity of explosive products behind the detonation front for a variety of explosives.[7]) Brish et al.[7] observed that a powdered TNT-hexogen (Comp B) explosive had a resistivity five times as high as the same explosive in molded form. A limited number of explosive conductivity experiments were performed by the authors using probe techniques similar to that of Cook.[6] The results obtained suggested that the resistivity profile behind the detonation front could be approximated as

$$\rho(t) = \rho_0 \exp(t/T) \tag{1}$$

For Comp B molded explosive, the constants are $\rho_0 = 0.5$ Ω-cm and $T = 1$ μs. In contrast, granular explosives, such as PETN and nitroguanadine, exhibited ρ_0 values an order of magnitude higher than the more homogeneous cast material in Comp B. Detasheet explosive exhibited a monotonically falling resistivity to a value

Table 1

Explosive	Detonation[3] Velocity (mm/μs)	Detonation[3] Pressure (kilobars)	Breakdown Field of Product Gas (kV/cm)	Form	Average[1] Resistivity (ohm-cm)	Soviet Equivalent
TNT	4.5	198		powdered	0.25	TNT
PETN	5.2	734	100–120[2]	powdered	7.6	PETN
COMP B	7.84	269		molded	0.2	50% TNT/
	5.7			powdered	1.0	50% Hexogen
Octol	8.5	309				
9404	8.84	337				
Nitroguanadine	5.5	75		powdered	> 10	
Lead Azide	2.7			powdered	0.16	Lead Azide
RDX	6.2		100–120[2]	powdered	4.3	Hexogen
---	5.0			powdered	6.3	Tetryl

1. Reference 7.
2. Reference 10.
3. Reference 14.
4. Reference 15.

less than 0.025 Ω-cm with no indication of rise within the observation time of 5 μs. Since the latter explosive can be expected to contain a significant fraction of unreacted carbon as a detonation product, it is assumed that "afterburn" of this unreacted carbon contributed to the reduction of resistivity.

Dielectric breakdown of the explosive product gases behind the detonation front has also been measured by several experimenters. In these experiments an external source was used to apply a high voltage to the detonation product gases, and the timing was adjusted to avoid the conductive region of the detonation front. PETN and RDX (hexogen) have been evaluated, and the breakdown strength was in the range of 100-120 kV/cm.[8,9,10,11] The pressure of the detonation products affected the breakdown strength, with the higher pressures giving the higher breakdown strength. Ershov et al.[12] also noted that the breakdown strength of PETN could be increased by 25% with the addition of a fluoroplastic. Datsenko et al[9] have studied the effect of metal electrodes on the breakdown strength of the explosive. They found that the breakdown strength was lower within a thin boundary layer (depth of a few mm) adjacent to the metal surface. Aluminum showed a much stronger effect than did copper or steel, and the authors inferred that ablation of the metal surfaces was responsible for the lower breakdown strength.

PLANAR SWITCHING EXPERIMENTS

Experimental study of the switching mechanisms of an explosive switch is more easily conducted in a planar geometry.[13] The plasma is more directly accessed; the uniformity of the explosive detonation front is better controlled; and the explosive material can be readily changed. A series of planar experiments are described in Ref. 16, and the experimental arrangement is shown in Fig. 2. A low inductance feed to the switch was obtained with a folded strip line that was connected to a capacitor bank current source. The plasma switching element was a planar slab, which was confined on one side by a dielectric tamper and on the other side by an explosive surface. An electrically-exploded aluminum film (operated at atmospheric pressure in air) was used to fill the cavity with plasma; the resultant plasma was predominantly nitrogen and oxygen, with about 2% aluminum.

Compression of the plasma channel was accomplished with a slab of explosive that was detonated by a plane-wave-generator. The detonation wave was planar within a few tens of ns, thus assuring uniform, symmetrical compression of the plasma channel. To eliminate the uncertainty about interaction with walls, channel compression was also produced by two opposed explosive charges. This arrangement also had the advantage of doubling the effective closure velocity of the explosive.

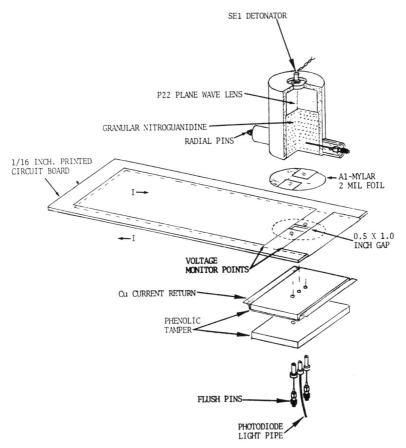

Fig. 2. Planar explosive opening switch.

These and complementary experiments at Los Alamos National Laboratory[5] show two regimes of operation for the explosive switch. A typical resistance profile is shown in Fig. 3. Results obtained in several experiments are tabulated in Table 2. The fast regime is typified by a resistance e-fold time, T_1, of 0.1-0.2 μs, and a peak resistance, R_2, of about 0.1 Ω. Beyond this peak, the resistance falls for 1 μs and then begins to rise again with an e-fold time, T_2, on the order of 1 μs. A comparison of this resistance profile with theoretical models will be discussed in the following section. In these low-current experiments, the switch will completely shut off the current source, and withstand voltage gradients of at least 20 kV/cm. This gradient does not appear to be the dielectric breakdown limit of the switch, since the peak voltage does not increase when the plasma channel length is increased.

Table 2. Typical Planar Switch Results

Shot No.	Explosive	Current (kA)	R_1 (mΩ)	R_2 (mΩ)	T_1 (ns)	R_3 (mΩ)	T_2 (μs)	V_1 (kV)	V_2 (kV)
1	COMP B	199	7	113	1200	87	4.26	19	14
2	COMP B	229	7	120	108	30	1.45	22	20
3	COMP B	212	7	158	167	43	1.35	33	52
4	COMP B	109	6	70	102	25	1.78	14	35
5	COMP B	173	17	181	139	48	1.31	28	41
6	Nitroguanadine (1.3 g/cc)	210	7	187	254	143	2.89	37	31

Shot 1: Single 1.27 cm thick COMP B pad fired into a 2.54 x 2.54 x .15 cm switch gap with a 0.15 cm thick tamper.

Shot 2: Opposed 1.27 cm thick COMP B pads fired into a 2.54 cm long x 5.08 cm wide x 0.33 cm thick switch cavity. 1000 μF, 600 nH, 9.5 kV capacitor discharge current source.

Shot 3: Opposed 1.27 cm thick COMP B pads fired into a 2.54 x 2.54 x 0.33 cm switch cavity. 1000 μF, 600 nH, 10 kV capacitor discharge.

Shot 4: Opposed 1.27 cm thick COMP B pads fired into a 7.62 cm long x 2.54 cm wide x 0.33 cm thick switch gap. 1000 μF, 600 nH, 10 kV capacitor discharge.

Shot 5: Opposed 1.27 cm thick COMP B pads fired into a 2.54 x 2.54 x 0.33 cm switch cavity. 1000 μF, 8.7 kV, inductance reduced to 300 nH.

Shot 6: Opposed, 2.54 cm, 1.3 g/cc, pressed Nitroguanadine pads fired into a 2.54 x 2.54 x 0.33 cm switch cavity. 1000 μF, 600 nH, 9.5 kV capacitor discharge.

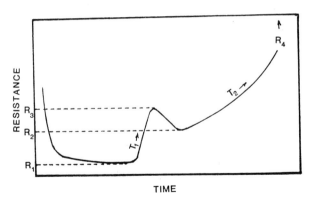

Fig. 3. Representative planar switch resistance profile. R_1, R_2, and R_3, and R_4 are switch resistance first minimum, first maximum, second minimum, and second maximum, respectively. T_1 and T_2 are the e-fold time constants of the first and second resistance.

THEORETICAL MODELING

The behavior of the Pavlovskii type explosive opening switch has been modeled by several investigators.[17,18,19] Typical of these models is that of Greene et al,[18] in an extension of the Raven 1-D hydrodynamic code, and that of Tucker[19] in a similar ideal gas analytical model. The geometry considered in both models is shown in Fig. 4. It can be seen that the system is simply the adiabatic compression of the gas in the arc channel, C_2, between the expanding explosive products, C_1, and the immovable tamper. Gases C_1 and C_2 are assumed to be separated by a thin flyer, M, typically utilized to initiate the arc. Neglecting the relatively small ohmic heating during channel compression, energy conservation yields.

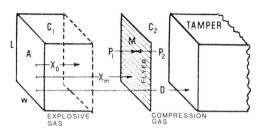

Fig. 4. Typical 1-D planar switch model.

$$E_c = C_1 e_1 + KE_1 + KE_m + C_2 e_2 + KE_2 \tag{2}$$

where KE_1, KE_m, and KE_2 are the kinetic energies of the explosive gas, flyer, and compressed plasma, respectively, and E_c, $C_1 e_1$, and $C_2 e_2$ are the explosive chemical energy and the instantaneous internal energies of the explosive gas driver and receptor plasma. It can be shown from simple ideal gas relationships that the differential equation of flyer motion defining the arc channel boundary is[19]

$$\ddot{X}_m = \frac{1}{M} \left\{ \left[E_c - \frac{\dot{X}_m^2}{2} \left(M + \frac{C_1}{3} + \frac{C_2}{3} \right) - \frac{P_{20}(D-X_0)^{\gamma_2}}{(\gamma_2-1)(D-X_m)^{\gamma_2 - 1}} \right] \right.$$

$$\left. \cdot \left[\frac{\gamma_1 - 1}{X_m} \right] - P_{20} \left[\frac{D-X_0}{D-X_m} \right]^{\gamma_2} \right\} \tag{3}$$

where M, C_1, and C_2 are the mass/area of flyer, driver, and receptor gases; E_c, the explosive chemical energy per unit area; P_{20} is the initial pressure of the heated arc channel; X_m is the instantaneous flyer position (between X_0 and the tamper position D); and γ_1 and γ_2 are gas constants of the explosive gas and arc channel. Equation (3) may be evaluated numerically, yielding a time dependent solution of the flyer displacement and thus compression of the gas, C_2. From this displacement of the compressed gas temperature, T_2, and pressure, P_2, are given by

$$T_2 = T_{20} \left(\frac{X_0}{X_m} \right)^{\gamma_2 - 1} \tag{4}$$

and

$$P_2 = P_{20} \left(\frac{X_0}{X_m} \right)^{\gamma_2}, \tag{5}$$

respectively.

Finally, if it is assumed that electrical conductivity is given by the Spitzer relationship[20] i.e., $(\sigma \sim T_2^{-3/2})$, time dependent resistivity and resistance profiles may be straightforwardly derived. Examples of solutions of Greene's MHD model[18] and the ideal gas model are shown in Figs. 5 and 6. It can be seen that very similar oscillatory solutions are predicted by both models.

The associated resistance profile of this predicted compression-bounce-recompression solution is compared in Fig. 7 to that experimentally observed in a planar test device. It can be seen that although the initial resistance excursion is approximated by the simple channel compression, the profiles differ markedly at later times. In contrast to the predicted repetitive resistance behavior, the experimental waveform exhibits a relatively slow exponential rise following the initial compression spike.

The mechanism of this slow, microsecond, resistance rise is not well understood; however, it is likely that the increase is the result of cooling of the arc channel, possibly by heat transfer to the relatively cool explosive gas and a reduced ohmic heating resulting from a generally falling circuit current associated with the high arc channel resistance. It has been suggested in the model developed by Baker[17] that cooling results from Rayleigh-Taylor instability and the associated turbulent mixing of the relatively cool explosive products into the arc channel.

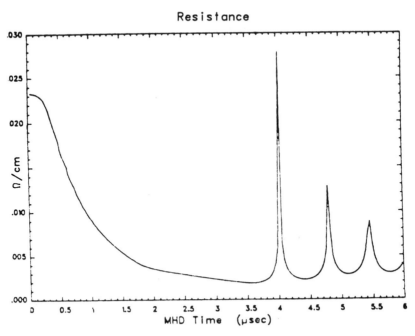

Fig. 5. MHD solutions of planar switch model.

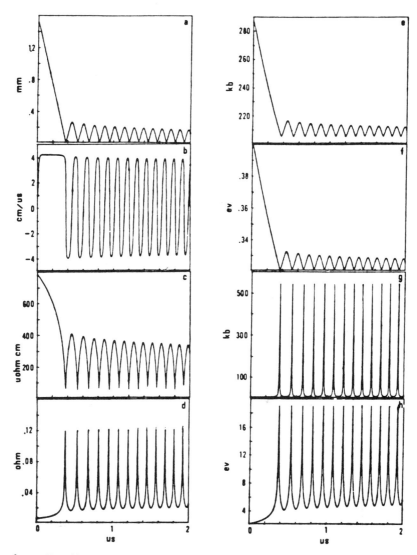

Fig. 6. Predicted parameter profiles from Ideal gas model. (a) Gas thickness; (b) flyer velocity; (c) arc resistivity; (d) arc resistance; (e) explosive gas pressure; (f) explosive gas temperature; (g) compression gas pressure; (h) compression gas temperature.

Recent 2-D modeling work by Greene and Bowers[21] support this hypothesis. They show that an initial perturbation at the plasma-explosive boundary indeed becomes Rayleigh-Taylor unstable at the point of maximum compression of the plasma channel. If this instability leads to complete mixing of the relatively cool explosive

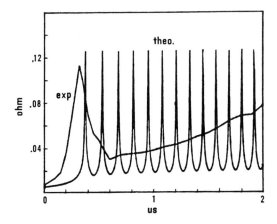

Fig. 7. Comparison of theoretical and experimental resistance profiles.

products with the hotter plasma, then the plasma resistivity will increase by about a factor of 100. If only half of the perturbed explosive mass mixes, then the plasma resistivity will increase by a factor of 70. The actual fractional mixing is not known, but it appears reasonable to expect a factor of 10 or more resistivity increase from this effect.[21]

In conclusion, models developed to date, based upon simple 1-D gas behavior, can adequately describe only the initial compressive stage, producing roughly a 20-fold increase in arc resistance. The second phase of the switching profile depends upon additional processes such as channel cooling by turbulent mixing with explosive reaction products. Further analysis of these more complex 2-D mechanisms, including electromagnetic effects, is now required.

HIGH-CURRENT SCALING EXPERIMENTS

Large-current scaling experiments have been conducted with the cylindrical geometry originally proposed by Pavlovskii. This geometry has the advantage of low inductance, high current density, and ease of scaling to high energy applications. As mentioned earlier, this geometry has the disadvantage of being less accessible for plasma measurements during the switching process and presenting a more difficult problem with explosive detonation symmetry. References 4, 16, and 22 give results from these experiments with two basic switches, one with a plasma channel of 4.6 cm diameter and 2.5 cm height and the other with 20 cm diameter and 10 cm height (Fig. 8). Some of the tests were driven with a capacitor bank, while others were driven by explosive generators. The

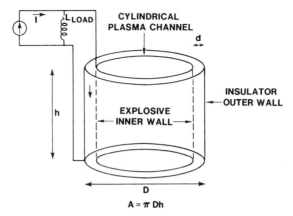

Fig. 8. Experimental arrangement for scaling experiments.

current through the switch was varied from 0.85 to 3.7 MA, with the linear current density (switch current divided by plasma channel circumference) varying from 30 to 180 kA/cm. Typical peak resistance values of 100 mΩ were obtained, as seen from the current, voltage, and resistance data set in Fig. 9. These data indicated that the resistive e-folding time was related to the square of the current density. This scaling result can be understood with the following dimensional analysis.

Consider an equilibrium model of the plasma in which the increased resistance of the switch occurs as a result of energy losses from the plasma surfaces during compression. The electrical power dissipated resistively in the plasma is removed from the surface through a combination of radiative and turbulent convection processes that are represented dimensionally by the function e(A,t), where A is the surface area of the plasma and t is time. For the simplest case we can assume that the energy loss term is linearly related to the surface-to-volume ratio for the plasma. For this geometry the ratio is just the reciprocal of the annular plasma channel thickness d. Then this equilibrium model gives

$$RI^2 = e(A,t)A = e_o \frac{A}{d} . \tag{6}$$

Also, for a purely inductive energy storage

$$L\dot{I} + IR = 0 . \tag{7}$$

Integration of these two equations gives the solution

$$I(t) = I_o \sqrt{1 - t/T} \tag{8}$$

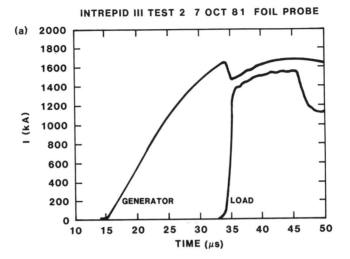

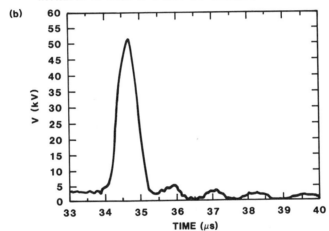

Fig. 9   Current, voltage, and resistance profiles for Intrepid II, Test #2. This test switch was driven by a capacitor bank.

and

$$R = R_o / (1 - t/T) \,, \tag{9}$$

where

$$T = \frac{LI^2 d}{2e_o A} \tag{10}$$

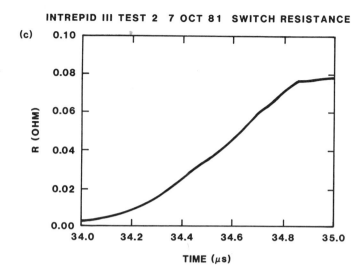

Fig. 9   Continued from previous page.

The resistance function (Eq. (9)) is similar in form to the exponential resistance rise observed experimentally. Consequently, the constant T should approximate the e-fold time of the resistance rise. Note that T (Eq. (10)) includes the current-density-squared dependence that was observed in the scaling experiments. Figure 10 is a plot of the measured e-folding time versus the scaling parameter $S = LI^2 d/A$. The channel thickness, d, is the initial thickness before compression. This graph has data from several experiments listed in the caption and covers a wide range of physical size and currents. The scaling equation is, in MKS units:

$$T = 7.8 \times 10^{-12} S \ . \tag{11}$$

Of particular importance is the conclusion that the explosive surface area, A (= hw), must increase as switch current increases to maintain a given switch e-folding time.

As an example of a high-energy experiment comparable to Pavlovskii's original work, Fig. 11 presents data from a test driven by an explosive generator and using the larger of the two cylindrical switches. As reported in Ref. 22, a 70 GW power pulse was transferred to a 20 nH inductive load in 1 µs. The inductance of the switch was 63 nH. The switch e-folding time was 0.5 µs, the switch current, I, was 3.2 MA, and the explosive area, A, was 630 cm^2. This experiment was driven with an explosive generator which delivered 260 kJ to the switch. A total energy of 60 kJ was transferred to the inductive load for an efficiency of about 20%. The theoretical transfer efficiency from an inductive store to an inductive load is

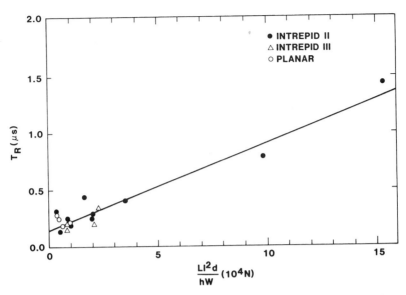

Fig. 10. The resistance e-fold time scale linearly with the parameter $S = LI^2d/hw$. T_R is in microseconds and S is in MKS. Data are tabulated in the Appendix.

$$\epsilon = \frac{L_1 L_2}{(L_1 + L_2)^2} \qquad (12)$$

which in this case is 18%. Therefore, the experimental transfer efficiency for this switch is at the theoretical value for this combination of inductances.

Experiments have shown that proper operation of the switch is dependent on the formation of a uniform plasma.[4] If the plasma is formed with a non-uniform, longitudinally-channeled structure, as depicted in Fig. 12, the switch will open slowly and peak at a low resistance. Figure 12a shows a streak photograph of a uniform plasma during switching. In this experiment, the plasma channel is observed through a transparent outer insulator made of lucite. The compression of the plasma channel is marked by an increase in light output as recorded by the narrow, bright band in the streak record. Mixing of the explosive product with the hot plasma follows, and the light intensity drops quickly until the plastic insulator becomes opaque from the incident shock wave. The variation of the compression wave arrival time across the channel is shown by the wavy character of the bright compression line. The non-simultaneity of the compression wave arrival is a few tenths of a microsecond. The resistance profile of the switch is plotted below the streak record with the proper time registration. In Fig. 12b the same observations are made, but here the switch performance was much poorer, with a slower rise and much lower peak resistance. A series of such experiments led to the conclusion that the poor switching resulted from the formation of discrete, narrow channels in the plasma, as sketched in Fig. 12b. The plasma then is not compressed uniformly by the explosive but tends to be merely deflected around these higher density regions. Evidence of this is seen in the streak record in Fig. 12b, with long "fingers" of hot plasma extruded between the regions which are cooled by mixing with the explosive products. These channels are formed in the initiation process of the plasma and have been observed to endure for at least 20 μs. The longitudinal channels were found to form when the initiating current derivative per unit circumferential width was below 6×10^8 A/cm·s. When the current density derivative was above this level, good switch performance was consistently obtained with the 300 Å aluminum films used in these tests. For operation from a capacitor bank this current derivative is readily exceeded, but it does present a design constraint for operation with an explosive generator, which begins with a large initial inductance and consequently a low initial current derivative. If necessary, the initial current derivative can be increased with a peaking inductance and closing switch, as was done for many of these explosive generator tests.

FUTURE RESEARCH DIRECTIONS

Pavlovskii's experiments and other more recent work have shown that the explosive opening switch can provide fast switching of large currents with good energy transfer efficiency. High power switching, at 0.4 TW, has been demonstrated by Pavlovskii. Because of the large amount of explosive involved in the switch, it is by nature a destructive device and is most readily incorporated into single-shot applications. Thus the switch naturally goes with explosive generators, both from an operational as well as electrical point of view.

The scaling relation in Eq. (11) predicts that the surface area between the plasma and explosive must increase as current, and energy, are increased. The large coaxial experiments in Refs. 3 and 4 had surface areas of approximately 1000 cm^2, load energies of about 100-200 kJ, and transfer times of about 0.5 μs. When this output is scaled to 1 MJ and 0.5 μs, the surface area must be about 3200 cm^2. Therefore a fast, high energy experiment will require a very large explosive surface area and may well require more explosive than the generator which is driving the system. Although the scaling relations show that this switch can perform at the terrawatt power level, it is not apparent whether the switch would be practically useful for operational reasons. Perhaps a

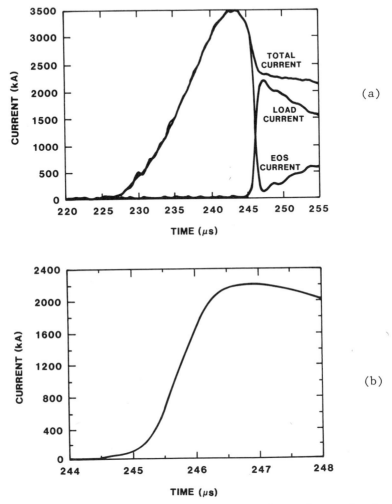

Fig. 11. Experimentally observed outputs from the explosive generator-cylindrical explosive opening switch system.
(a) system currents
(b) load current expanded.

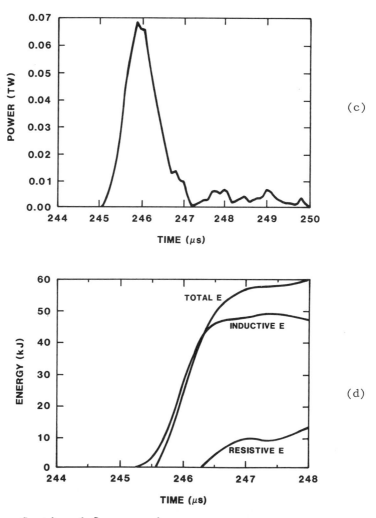

Fig. 11. Continued from previous page.
Experimentally observed outputs from the explosive generator-cylindrical explosive opening switch.
(c) load power
(d) load energy.

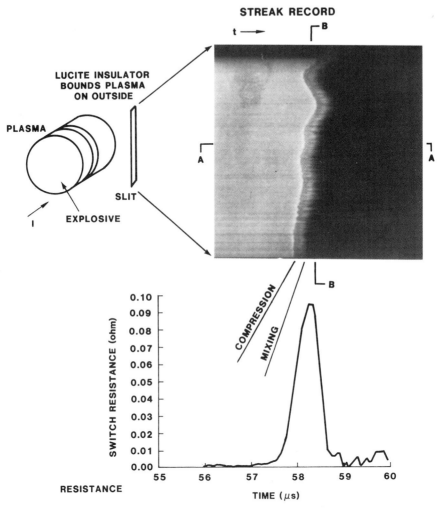

Fig. 12a. Streak camera record of the interaction between the switch plasma and explosive product gases. Camera slit is oriented across the equator of the cylindrical plasma, plasma, looking through the lucite insulator.

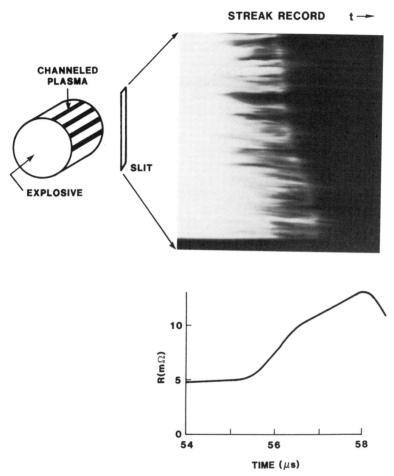

Fig. 12b. Streak photography of plasma/explosive interaction with a strongly channeled plasma. Resistance rise is slower than that from a uniform plasma, and peak resistance is lower.

hybrid combination of explosive generator and switch can be designed to minimize the total amount of explosive required and thus make handling and operational requirements easier.

Further studies of explosive product electrical characteristics are needed to optimize switch resistivity and breakdown characteristics. The ideal explosive must give fast opening time, high peak resistance, and high voltage stand-off. This ideal explosive has probably not been identified as of yet.

Finally, the 1-10 μs time regime seems to be an area where the explosive opening switch can give very good performance in a compact package. This regime has not been completely explored, but the very high resistance and high current density that can be reached with a few microsecond rise time holds promise for improved switching capability.

REFERENCES

1. R.D. Ford and I.M. Vitkovitsky, High Recovery Voltage Switch for Interruption of Large Currents, Rev. Sci. Inst., 53:1098 (1982).
2. P.S. Levi, J.D. Watson, M.L. Hines, and R.E. Reinovsky, Staged Explosively Driven Opening Switch Development for Explosive Flux Compression Generators, Proceedings of the 4th IEEE Pulsed Power Conference, Albuquerque, NM (1983).
3. A.I. Pavlovskii, V.A. Vasyukov, and A.S. Russkov, Magneto-implosive Generators for Rapid-risetime Magampere Pulses, Sov. Tech. Phys. Lett., 3:320 (1977).
4. B.N. Turman, T.J. Tucker, and P.J. Skogmo, Megampere Current Experiments with the Intrepid Fast Opening Switch, Sandia National Laboratories Report SAND82-1531 (1982).
5. J.H. Goforth and R.S. Caird, Experimental Investigation of Explosive-Driven Plasma Compression Switches, 4th IEEE Pulsed Power Conference, Albuquerque, NM (1983).
6. M.A. Cook, "The Science of High Explosives," Reinhold Publishing Co., New York, 143-171 (1958).
7. A.A. Brish, M.S. Tarawov, and V.A. Tsukerman, Electric Conductivity of the Explosion Products of Condensed Explosives, Sov. Phys. JETP, 37:1095 (1960).
8. V.L. Korolkov, M.A. Melnikov, and A.P. Tsyplenko, Dielectric Breakdown of Detonation Products, Sov. Phys. Tech. Phys., 19:1569 (1975).
9. V.A. Datsenko, V.L. Korolkov, I.L. Krasnov, and M.A. Melnikov, Dielectric Breakdown of Detonation Products, Sov. Phys. Tech. Phys., 23:33 (1978).
10. P.I. Zubkov, L.A. Lukjanchikov, and Yu. V. Ryabinin, Electric Strength of Scattered Detonation Gases, J. Appl. Mech. and Tech. Phys., 17:109 (1976).

11. P.I. Zubkov, L.A. Lukjanchikov, and K.A. Ten, Quenching of a Developed Arc through Shock Compression by Explosion Products, <u>Zhurnal Prikladnoi Mekhaniki i Tekhnichskoi Fiziki</u>, 5:128 (1981).
12. A.P. Ershov, L.A. Lukjantshikov, Ju. V. Rjabinin, and P.I. Zubkov, Electrophysical Properties of Detonation Products of Condensed Explosives, in "Megagauss Physics and Technology," P.J. Turchi, ed., Plenum Press, New York (1980).
13. L.A. Lukjantshikov, K.A. Ten, and P.I. Zubkov, Extinguishing of the Electric Arc Compressed by Shock Waves, in "Megagauss Physics and Technology," P. J. Turchi, ed., Plenum Press, New York (1980).
14. R.L. Jameson and A. Hawkins, Shock Velocity Measurements in Inert Monitors Placed in Several Explosives, Preprints, the 5th Symposium on Detonation, <u>Office of Naval Research Report ONR DR-163</u>, 17 (1970).
15. H.C. Hornig, E.L. Lee, M. Finger, and J.E. Kurrle, Equation of Sate of Detonation Products, Preprints, 5th Symposium on Detonation, <u>Office of Naval Research Report ONR DR-163</u>, 422 (1970).
16. B.N. Turman and T.J. Tucker, Experiments with an Explosively Opened Plasma Switch, <u>Proceedings of ARO Workshop on Repetitive Opening Switches</u>, Tamarron, CO, DTIC NO. AD-A110770, M. Kristiansen and K. Schoenbach, eds, (1981).
17. L. Baker, Theoretical Model for Plasma Opening Switch, <u>Sandia National Laboratories Report SAND80-1178</u> (1980).
18. A.E. Greene, J.H. Brownell, T.A. Oliphant, G.H. Nickel, and D.L. Weiss, Computational Simulations of Explosive Driven Plasma-Quench Opening Switches, <u>4th IEEE Pulsed Power Conference</u>, Albuquerque, NM (1983).
19. T.J. Tucker, Simple Ideal Gas Model of the Pavlovskii High-Explosive Opening Switch, <u>Sandia National Laboratories Report SAND83-1325</u> (1983).
20. L. Spitzer, "Physics of Fully Ionized Gases," Wiley-Interscience, New York (1962).
21. A.E. Greene, and R.L. Bowers, Two-Dimensional Modeling of High Explosive Driven Plasma Opening Switches, J. Appl. Phys., 67:78 (1985).
22. T.J. Tucker, D.L. Hanson, and E.C. Cnare, Experimental Investigation of an Explosive-Driven Pulse Power System, <u>Sandia National laboratories Report SAND83-0180</u>, (1983).

DIELECTRIC BREAKDOWN IN SOLIDS

A.K. Jonscher

Physics Department
Royal Holloway and
Bedford New College
Egham, Surrey, TW20 OEX, UK

INTRODUCTION

This brief review of dielectric breakdown processes in solids is written with particular reference to short-time breakdown under impulse conditions and less emphasis will be placed on long-term degradation under high electric stress which is of more direct interest in electrical power industry. Nonetheless, the background discussion will be cast in fairly general terms, since the same principles apply to all types of breakdown once it ensues.

Breakdown in solids is not a simple phenomenon - it involves a range of processes which vary according to the particular material and the stressing conditions, with the influence of the electrodes a very important parameter. The basic understanding of the various elementary physico-chemical phenomena involved in breakdown is not well advanced, with the result that there exist many theories of which every one may be faulted on one ground or another.

Breakdown has to be taken as the resultant of these different elementary processes but it is very difficult to analyze it in terms of separate treatments of the various components which could then be re-assembled to form the complete picture. Such processes as carrier injection, their movement in the electric field, the resulting heating effects, defect formation in the material, avalanching, and other non-linear phenomena may be defined and analyzed individually, but their interpretation is too complex to enable a complete picture to be obtained from them.

A serious complication arises from the fact that the available experimental evidence on which to build theoretical models is generally very incomplete because of the need to collect considerable statistical information on any one of the many aspects of breakdown. Thus the requirement for a data base is formidable and there is little hope that an adequate experimental program can be mounted and the results be made available in suitable form for evaluation.

The approach adopted in the present review is, therefore, to attempt to present a balanced overview of the factors which in the Author's opinion are important in influencing the breakdown process and to indicate, where appropriate, the complications arising from the interactions between them. In this approach the emphasis is placed on outlining the broad trends, even at the expense of some inevitable oversimplification.

GENERAL FEATURES OF BREAKDOWN IN SOLIDS

An examination of the available experimental evidence enables certain general features of solid dielectric breakdown to be discerned and they may be summarized as follows.

- a) Breakdown has an essentially statistical nature and usually follows the Weibull distribution involving a field-time action integral which implies a "memory" of past history of stressing of the material[1-3].

- b) The ultimate result of breakdown is the thermal formation of a narrow channel of destruction joining the electrodes across the dielectric material.

- c) Breakdown may be preceded and accompanied by certain visible irreversible signs such as the various forms of "trees"[4,5].

- d) Under the combined action of electric field, temperature and possibly other factors like chemical attack, the material undergoes an aging process which is sometimes divided into "long" and "short" term aging and which leads to progressive reduction of the ability to withstand high electric stress.

- e) Electronic or ionic charge carriers appear to be essential to bring about electrical aging.

- f) The often invoked distributed processes of carrier generation, such as avalanching by impact ionization and the Poole-Frenkel effect[6] are unlikely to be the dominant causes of breakdown in the majority of technically important cases and other phenomena have to be invoked.

g) Many "good" insulators having very different physical properties, such as some polymers, silica, mica, etc., show a remarkably narrow range of ultimate breakdown fields in the range of 1 GV/m.

The following discussion is intended to elaborate on some of these points and in this we shall be making reference to a recent review[7].

THE STATISTICAL NATURE OF BREAKDOWN

Dielectric breakdown in any medium, solid, liquid, gaseous or even in vacuum is not a deterministic phenomenon which can be predicted with certainty, once the parameters of the material and of the stress have defined. Certain statistical predictions can be made of the probability that breakdown will occur under given conditions and it is generally accepted that Weibull statistics provides a satisfactory description of the distribution of the breakdown probability $F(E,t)$ under the action of a time-dependent field $E(t)$:

$$\ln[1 - F(E,t)] \propto - \int_0^t E^b t^{(a-1)} \, dt \qquad (1)$$

where a and b are parameters characteristic of a given material. The generalized form of the time-dependent field given in Eq. (1) is not typically used in practice, where measurements are normally made with step-function or ramp voltage waveforms. In the former case, with a step amplitude E, the constant failure probability contours correspond to $E^b t^a$ = const, suggesting the following relationship between the breakdown field and the time of exposure to that field:

$$E_b \propto t^{-k}, \qquad (2)$$

where $k = a/b < 1$, since normally a<b. This shows that the breakdown field decreases with the time of application of stress following the power law, Eq. (2), and this is typically observed in practice, although the exact form of this relationship is difficult to ascertain, especially in view of the limited number of experimental data and of their inevitable scatter.

Although Weibull statistics are reasonably well borne out by experimental data where they are available in sufficient quantity, there is no evident theoretical explanation of its applicability. Dissado and Hill have proposed[8] that Weibull statistics may be a natural consequence of the fluctuations which are expected in the local field in a dielectric material of which the relaxation process is governed by their many-body theory of dipolar interactions[9]. This prediction, if proved correct, would represent the first major success in explaining the nature of breakdown statistics. However, the applicability of this theory which presupposes an ideally homogeneous medium, to real systems in which there exist local inhomogeneities, remains to be explored further.

From a purely pragmatic point of view, the fundamental reasons for the applicability or otherwise of Weibull statistics are of lesser significance than the fact that a relationship of the type suggested by Eq. (2) is valid at all. This means that the dielectric strength decreases with time of exposure to stress. In the case of long-term exposure, e.g. in electrical insulation in cables and machinery, this decrease may be considerable and this constitutes the limitation of the useful life of insulation.

The conclusion from the point of view of pulse applications is that the dielectric strength is likely to be high, because of the very limited total exposure time to stress.

It should be stressed that the validity of the field-time dependence suggested by Eq. (2) implies the presence of some form of "memory" of past stressing conditions. This experimentally well established feature[7] constitutes the principal argument against deterministic theories of breakdown which depend on uniformly distributed processes that do not give rise to defect formation, such as avalanching. We shall return to this point later.

SOURCES OF CHARGE CARRIERS IN INSULATORS

In view of the established fact that charge carriers are necessary to produce breakdown[10,11] it is important to obtain a clear idea of the various ways in which they may be introduced into insulators.

Any solid contains a finite density per unit volume of more or less slowly mobile electronic or ionic charge carriers. Their density is dictated by thermodynamics and by the presence of suitable impurities or defects which provide sites where "free" carriers may be liberated with a relatively small input of energy. This process is ultimately one of semiconductor action and, while it cannot be eliminated completely, its extent and the resulting rate of production of charge carriers in the system may be influenced by the control of impurities and imperfections.

In general, good insulators do not have significant densities of thermal equilibrium carriers and breakdown processes have to rely instead on the introduction of non-equilibrium carriers into the system under the action of an applied electric field. The most obvious source of such carriers is normally the electrode and it should be stressed that the removal or injection of carriers from an insulator by an electrode is the inevitable consequence of the difference of work functions between insulator and the metal[12,13] and does not, as such, require the presence of a high electric field.

The presence of a high field is important in two respects. Firstly, the injection of electrons from the electrode may be very considerably enhanced in this way, especially if field concentration arises from some inhomogeneity such as sharp point on the electrode - insulator interface. The resulting strongly divergent electric field may exceed the mean field by an order of magnitude or more and its effects can be very serious. Considerable care is taken in practical systems to avoid this type of imperfection and one solution is the use of non-metallic electrodes, such as semi-conducting rubber, in which there is less tendency to form sharp points and their effect is less serious because of the high resistivity of the material. The second effect of the field on charge carriers is the production of transport processes which will be discussed later.

In addition to electrode induced charge injection, charge carriers may be introduced into the volume of the material by volume generation at structural defects and imperfections, especially if these give rise to strongly divergent fields, as in the case of metallic particles in the volume of the insulator. This manner of charge formation is independent of the nature, or indeed the presence of electrodes - for instance in the case of laser damage to dielectrics this is the main cause of generation of carriers, although quantum effects may become significant in liberating electrons by bond breaking.

It is very difficult to quantify the phenomena of carrier production in the volume of materials because the exact mechanisms responsible are not clearly understood and, in any case, they probably differ between one material and another. One favorite theory being invoked is the impact avalanching due to energetic electrons gaining energy from the field and giving it up to the lattice with liberation of more free carriers. This type of mechanism is well known and understood in crystalline materials in which electronic mobilities are relatively high as a result of high crystalline perfection and of strong bonds in three dimensions, which favors wide energy bands for electronic states. It is likely that the breakdown process in thermal SiO_2 in Metal-Oxide-Semiconductor (MOS) structures may be of this type[14,15]. Claims are also being advanced that this may be applicable in other cases[16] but these explanations should be treated with caution in terms of what actually determines breakdown in practically important cases.

An essential feature of the avalanching process is the ability of electrons to develop a sufficiently long free path without energy-dissipating collisions for them to acquire, in the prevailing field, an energy sufficient to disrupt the bonds, i.e. several electron-volts. This mean free path is related to the low-field mobility, μ, in the sense that a material with high μ also has a long mean free path, but the converse is not necessarily true for high-energy electrons. It is possible to have a material with low or very low μ at low fields, due to trapping or defect scattering, and yet once a carrier has reached an energy of 0.2 - 0.4 eV its movement may be much less hindered.

The criterion is not so much the disorder as the nature of the energy bands in the solid in question. While the three-dimensionally strongly bonded solids have wide energy bands for electron motion in all directions, the situation in polymeric solids with their one-dimensional strong bonds is much more restrictive from the point of view of the freedom of electrons to gain high energies in elevated electric fields.

One feature of solid insulators which is especially likely to affect the behavior in high electric fields is the presence of internal "voids" of microscopic dimensions. These may lead to the formation of discharges with the consequent emission of secondary electrons and the resulting structural damage. These so-called partial discharges are especially important under transient and alternating electric fields, since, with steady bias of one polarity, these voids become ineffective once they have been polarized by an incipient discharge. They may, however be relevant to breakdown under repetitive impulses.

Our discussion of carrier injection into solids was entirely concerned with electrons. The reason for this is that, as a general rule, it is very difficult to introduce ionic charges into a solid from external electrodes. With highly specialized ion-selective electrodes it may be possible to introduce, for example, protons or sodium but ordinary metallic electrodes are not well suited for this purpose. It is our opinion that ionic charge carriers are effectively introduced only by the processes of ionization of lattice atoms in the volume of the material and electrode injection is not an important process.

TRANSPORT OF CHARGE CARRIERS IN INSULATORS

Once electrons or ions have been introduced into a solid, their subsequent movement in electric fields is determined by their mobility, μ, which defines the drift velocity, v,

$$v = \mu E , \qquad (3)$$

DIELECTRIC BREAKDOWN IN SOLIDS

where μ for electrons ranges from 1 m^2/Vs in the highest-mobility crystalline semiconductors, to 0.01 m^2/Vs in diamond or Al$_2$O$_3$ and down to several orders below this in polymers and also in heavily disordered solids. Ionic μ is altogether lower by several orders of magnitude than electronic μ in view of the much more difficult transport of ions through solids. In the case of polymers, in particular, it appears that ionic μ cannot be determined from the otherwise known values of diffusion coefficients of neutral gaseous species, on account of the easy attachment of ions to the molecular chains[7].

Relationships between carrier transport and the various types of bonds in crystalline and amorphous solids are represented schematically in Table 1, where the distinction is made between 3-dimensionally strongly bonded, 2-dimensionally and 1-dimensionally bonded solids. Depending on the type of bonding, and the crystallographic order, the various transport properties are indicated for electrons and for ions.

As has been mentioned above, the values of low-field μ defined by the linear relation, Eq. (3), are not necessarily relevant to breakdown and pre-breakdown conditions in solids, since at sufficiently high electric fields electrons may not lose through collisions all the energy they gain from the field and their effective "temperature" may rise significantly above the lattice temperature. Such "hot" electrons may then have much higher mean free paths than "cold" electrons and this may lead to an impact ionization avalanche, as described above.

Such considerations would not be applicable to ionic transport, where the only way in which ions may acquire significant energies from the field is truly free-space flight, as for example in voids. Ions cannot gain high energies in a solid in the manner comparable to electrons, since they cannot propagate as quasi-free particles by quantum mechanical tunneling.

The commonly accepted transport theories of electronic and ionic charges in insulators are based essentially on concepts derived directly from the corresponding semiconductor phenomena. The inadequacy of such arguments in the context of high-field behavior of insulators was pointed out in Ref. 7, where the need was stressed to link charge movements in high fields to the progressive creation of lattice damage in the material. This process of "conditioning" of the insulator by the passage of charge carriers, whether electrons or ions, arises directly from the observed Weibull statistical relations in which the breakdown behavior is related to the time of action of a high field on the material, and therefore on the charge carriers.

Table 1

Classification of Solids According to Type of Bond and Charge Transport

Dimensionality of strong bonds	Examples of structures		Passage of ionic charges	Transport of electrons in the conduction band
	Crystalline	Amorphous		
3-D isotropic	Most oxides, sulphides, alkali halides	Inorganic glasses Amorphous semiconductors	Equally difficult in all directions	Equally easy in all directions
2-D layer structures	Mica boron nitride beta alumina		Difficult across planes - only through defects relatively easy along the planes	Difficult across the planes Easy along the planes
1-D chain structures	Certain 1-D lattices, crystalline polymers, spherulites	Amorphous polymers Rubbery above T_g glassy below T_g	Relatively easy in all directions, across the strong bonds or between them. In amorphous systems randomness enhances mobility	Equally difficult in all directions
0-D Wholly van der Waals bonds	Rare gas solids, "small molecule" solids	liquids	Relatively easy in all directions	Equally difficult in all directions

A second point which is also essential for any realistic model of breakdown in solids is the process of producing a constriction of flow from the initially more or less uniformly distributed transport in an uniform material to the final narrow destruction channel of the catastrophic breakdown. Fundamentally, a constriction may arise in one of two ways. It has been recognized for a long time that, given a "semiconductor-like" conductivity - temperature relation, once a current begins to flow Joule heating follows inevitably and this may give rise to increasing conductivity, which produces more current and even more heating, and so on. In this way, a run-away situation may develop locally if, in the initially uniform situation, some local fluctuation of temperature should lead to an instantaneous constriction of the current path. These are the considerations underlying the various theories of thermal breakdown[17,18]. We are of the opinion[7] that while these models may be applicable in some semiconductors, for example switching chalcogenide glasses[19], they are unlikely to be relevant to typical breakdown situations in insulators. One argument in favor of our proposition is that the magnitude of the breakdown field is very insensitive to the temperature of the material, especially at and below room temperature, and it is also very insensitive to the value of thermal diffusivity which would be expected to be a relevant parameter in thermal processes. Furthermore it is not at all clear how thermal breakdown phenomena should be leading to Weibull statistics of breakdown. In the specific context of impulse breakdown, it would appear that thermal processes are even less likely to be applicable, since they require a definite time to develop. This is not to deny that ultimate phases of any breakdown process must the thermal in nature, but our main concern here is with the stages that lead to the pre-breakdown phase.

The central point of the approach adopted in Ref. 7 is the notion that, with the prevailing values of mobilities and mean free paths expected in typical insulators, charge carriers do not gain sufficient energy from the field to cause major damage leading to the disruption of the lattice. It is proposed, instead, that movement of not very energetic carriers may nevertheless give rise to progressive extension of some local defects, that are inevitably present in any material. It is proposed that the extension of such defects may be effected with less energy than would be required for their formation in the first place in a "perfect" lattice. In this way, continuing passage of current through an insulator results in progressive enlargement of initially present point defects. While the initial distribution of such defects must be random in space, with increasing density and size of these defects the probability increases of several defects combining together to interact with one another. As such defect "clusters" increase in size, so the mobility of charge carriers within them also increases, enabling these carriers to gain more energy from the prevailing field, thus producing even more damage, and so on. The ultimate outcome of this process is the formation of extended macroscopic damaged regions and it is then a matter of statistical chance which combination of such defects becomes the first path of high current density leading to the formation of the thermally destructive breakdown channel.

A schematic representation, taken from Ref. 7, of these processes is shown in Fig. 1 where the successive "frames" from left to right indicate the successive stages of conditioning of the material. On the left is the situation in the absence of field, or with field below a threshold value, where point defects exist in equilibrium with the prevailing temperature. The application of a field in excess of a threshold value leads to the production of more defects and the enlargement of existing ones, with the generation of electronic and ionic charge carriers. In the middle frame,

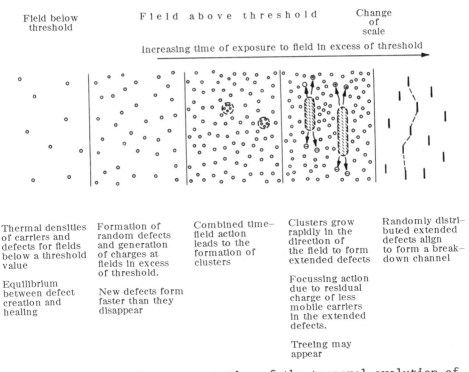

Fig. 1. A schematic representation of the temporal evolution of defect formation under the action of an electric field, leading to the ultimate breakdown of the dielectric material. The successive frames from left to right indicate successive stages in the process of conditioning of the material.

the critical stage has been reached where neighboring defects merge into clusters within which the carriers have a relatively higher mobility, so that they can acquire higher energies. This leads to a rapid growth of extended defects along the direction of the field - a process aided by the "focusing action" of the less mobile charge left behind by the more mobile carriers, constraining these to flow in preferred paths. The extreme frame on the right indicates, to a different scale, the manner in which a catastrophic channel may be formed from a series of extended defects.

The qualitative picture presented above does not lead us to any specific orders of magnitude of the critical fields and times which are required to produce the statistically predicted breakdown process. No such predictions can be made on the basis of any of the existing theories of breakdown, with the exception of Fröhlich's theory[20] applicable to alkali halides which represent a special case not typical of insulating materials, since they are crystalline materials with not very large energy gaps and high electronic mobilities.

It is evident from Eq. (2) that the experimentally observed close interrelation between time-to-breakdown and the value of the breakdown field points to the possibility of these times becoming very short in sufficiently high fields. Thus, it is proposed that very similar considerations to those outlined above apply to impulse breakdown with very short times, provided only that the fields are sufficiently high.

The interaction between a dielectric medium, crystalline or otherwise, and electronic or ionic charges moving through it is complex and must involve cooperative phenomena[9] of the same type which are responsible for low-field dielectric relaxation. Our understanding of these interactions, especially with reference to the high-field phenomena of interest in the present context, is far from complete. An early attempt to deal with this type of situation may be found in a paper by Fröhlich and Platzmann[21]. One important consideration is the storage of energy in the system following the passage of a charge carrier. In Ref. 7 it was suggested that such storage may be taking place between the passage of one carrier and the next on a given site, with the consequent accumulation of energy from particles, none of which, individually, had sufficient energy to cause lattice damage, leading to the formation of damage. This situation would be particularly relevant in circumstances in which several carriers are likely to follow in the same path at relatively short intervals of time.

Some idea of the energy relations arising from the movement of carriers in high electric fields in the presence of a mean free path ℓ between energy dissipating collisions may be obtained from Fig. 2. The field is plotted logarithmically as abscissa and the

mean free path vertically as ordinate, with constant energy gain
contours, W = eℓE, indicated. The various critical energy thresholds are indicated and the strengths of the various types of solid
bonds are shown. Van der Waals forces are typically between 0.01
and 0.1 eV, while the strong covalent bonds are around 10 eV. The
"intrinsic" breakdown strength is shown centered on 1 GV/m, as
discussed in the next section, the practical breakdown strength is
a factor of 10 below that and design strengths are a further decade
lower. On the ordinate scale the typical interatomic spacing is
shown as the lower limit to meaningful free path, ℓ, while the
sizes of microvoids or free volume regions are also shown. Impact
ionization and secondary electron emission processes at internal
void surfaces are shown in the region 10 - 20 eV. It is clear that
to achieve breakdown at 100 MV/m an effective free path of 0.1 μm
is required and this cannot be achieved readily in the absence of
extended defects in the material.

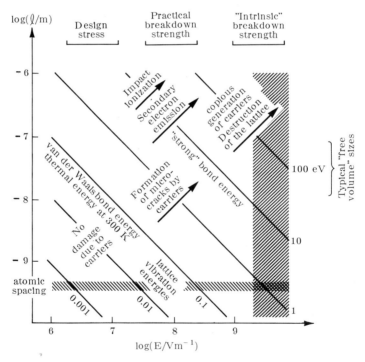

Fig. 2. Relationship between the electric field, E, the mean free
path of charge carriers in insulators, ℓ, and the energy
acquired by carriers in a mean free path. The vertical
shaded regions correspond to typical maximum values of
the breakdown field as shown in Fig. 3.

DIELECTRIC BREAKDOWN IN SOLIDS

THE MAGNITUDE OF THE BREAKDOWN FIELD

An overview of the magnitude of the breakdown field in typical insulators is shown in Fig. 3 which plots the field logarithmically and which also gives the values of the thermal diffusivity $\alpha = \kappa/C_p\rho$ for the various classes of materials. Here κ is the thermal conductivity, C_p the specific heat, and ρ the density. The point to note is that there is no apparent relationship between the breakdown field and the value of α which is relevant to thermal theories of breakdown, as mentioned above. The breakdown fields quoted in Fig. 3 are the highest recorded values with thin specimens and short breakdown times - it being understood that there is no absolute criterion of "intrinsic" breakdown.

The remarkable point appearing from Figure 3 is that the maximum breakdown field in a range of different "good" insulators reaches 1 GV/m, plus or minus a factor of approximately three. Since the physical and chemical properties of materials such as mica, silica and some polymers are so very different, it was suggested in Ref. 7 that the critical process may be tunnelling of electrons through a Schottky-lowered work function barrier - the Fowler-Nordheim process - which reaches massive proportions at fields of this order of magnitude. Once such copious emission does take place, there is no way in which breakdown can be averted because of the destructive effect of free charges at such high

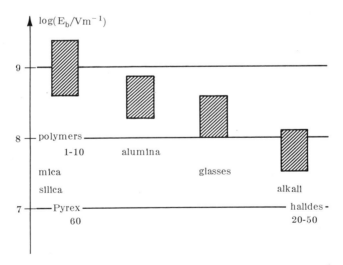

Fig. 3.   A logarithmic plot of the highest recorded values of the breakdown electric field, E_b, for a range of materials. The numbers indicate the magnitude of the thermal diffusivity.

fields on any lattice. It is also possible that the same basic process may be responsible for internal dissociation of electron-ion pairs in the volume of the material, with very similar effects. This would explain the apparent upper limit to breakdown fields.

BREAKDOWN UNDER SHORT IMPULSES

To these various considerations must now be added points of specific relevance to "short" pulse breakdown, the term being arbitrarily interpreted as implying pulse durations of the order of 10 - 100 ns with correspondingly shorter risetimes.

It was already mentioned that one relevant parameter in this situation is the total duration of electrical stress, which means the integrated time duration of a large number of pulses. Unless the trains of pulses involve very large numbers which add up to significant duration of stress, the implication is that the breakdown fields encountered with short pulses are likely to be near the upper limit expected for "short" delay times, i.e. of the order of 1 GV/m.

To these considerations, which represent a natural extension of fairly "classical" arguments may now be added a different process which is based on rather unfamiliar phenomena involved with rapid charging and discharging of insulators. It was shown recently[22,23] that, in complete contrast with alternating-current energy loss which is expressed by the familiar dielectric loss, $\epsilon"$ or $\tan \delta$, both of which may be functions of frequency, energy loss under impulse charging represents a relatively unknown subject. It was pointed out that, starting with the dielectric loss spectrum of the material, it is possible to derive expressions for the energy loss arising from sudden charging and discharging of a capacitor and that these losses may become significant in relation to the total stored energy. Under these conditions, it would be possible to obtain a significant rise in the temperature of the dielectric and this might contribute to a reduction of the breakdown voltage.

There is very little experimental material in the literature relating specifically to pulse breakdown. As a result, it is impossible to make an analysis of the limitations of the conventional approach when applied to these conditions. In particular, one question which needs to be answered is the extent to which the relationship between time-to-breakdown and the breakdown field, Eq. (2), can be extrapolated to very short times. The physical problem here is whether the aging processes which are responsible for medium-and long-term deterioration of the material, remain valid at very short times. It is intuitively evident that certain structural changes which form part of the aging process cannot be accelerated indefinitely under the action of sufficiently high electric fields. If this were so, then the breakdown voltage would eventually saturate below a certain pulse duration. At the present time, we do not appear to know the answer to that question.

REFERENCES

1. W. Weibull, A Statistical Distribution Function of Wide Applicability, J. Appl. Mech., 18:293 (1951).
2. E. Occini, A Statistical Approach to the Discussion of the Dielectric Strength in Electric Cables, IEEE Trans. Paper 71, TP 157 PWR (1971).
3. R. Kaneko and K. Sugiyama, Statistical Considerations on Impulse Breakdown Characteristics of Cross-Linked Polyethylene Insulated Cables, IEEE Trans. Power Apparatus and Syst., PAS-94:367 (1975).
4. P. Fischer, K.W. Nissen and P. Röhl, Elektrische Festigkeit von Polyethylen in Anwesenheit feldverzerrender Einschlüsse, Siemens Forsch.-u.Entwichkl.-Ber., 10:222 (1981).
5. R.M. Eichhorn, Treeing in Solid Extruded Electrical Insulation, IEEE Trans. Elect. Ins. EI-12: 2 (1976).
6. J.J. O'Dwyer, "The Theory of Dielectric Breakdown in Solids," Oxford University Press, London (1964).
7. A.K. Jonscher and R. Lacoste, On a Cumulative Model of Dielectric Breakdown in Solids, IEEE Trans. Elect. Ins., EI-19:567 (1984).
8. R.M. Hill and L.A. Dissado, Examination of the Statistics of Dielectric Breakdown, J. Phys. C: Solid State Phys., 16:4447 (1983).
9. L.A. Dissado and R.M. Hill, A Cluster Approach to the Structure of Imperfect Materials and Their Relaxation Spectroscopy, Proc. Roy. Soc., A390:131 (1983).
10. R. Fournie, Private communication.
11. T. Daykin, Theory of Voltage Endurance, 1983 Annual Rep. Conf. El. Insul. and Diel. Phenomena, IEEE Insulation Soc. (1983).
12. C. Barnes, P.G. Lederer, T.J. Lewis and R. Toomer, Electron and Ions Transfer Processes at Insulator Surfaces, J. Electrostatics, 10:107 (1981).
13. P.G. Lederer, T.J. Lewis and R. Toomer, Transport Processes in Polymeric Solids - the Role of the Surface and of the Electrodes. Annual Rep. Conf. El. Insul. and Diel. Phenomena, 58, IEEE Insulation Soc. (1980).
14. E. Harari, Conduction and Trapping of Electrons in Highly Stressed Ultrathin Films of Thermal SiO_2, Appl. Phys. Letters 30:601 (1977).
15. E. Harari, Dielectric Breakdown in Electrically Stressed Thin Films of Thermal SiO_2, J. Appl. Phys., 49:2478 (1978).

16. T. Hibma, H. R. Zeller, P. Pfluger, and Th. Taumann, A Model for Space Charge Injection in Dielectrics, <u>Annual Rep. Conf. El. Insul. and Diel. Phenomena</u>, IEEE EI, 259-265 (1985).
17. R. Goffaux, Sur les Mecanismes Physiques Responsables de la Disrupture d'Origine Thermique d'Isolants Solides Inhomogenes et Minces, <u>Bull. Sci. A. I. M.</u>, 93:71 (1980).
18. R. Goffaux and R. Coelho, Sur la Rupture Thermique Filamentaire Differee dans les Isolants Electriques, <u>Revue Phys. Appl.</u>, 16:67 (1981).
19. H. Fritzsche and S.R. Ovshinsky, "Amorphous and Liquid Semiconductors, Conduction and Switching Phenomena in Covalent Alloy Semiconductors," N.F. Mott North-Holland, (1970).
20. H. Fröhlich, Theory of Electrical Breakdown in Ionic Crystals, <u>Proc. Roy. Soc.</u>, A160:230 (1937).
21. H. Fröhlich and R.L. Platzmann, Energy Loss of Moving Electrons in Dipolar Relaxation, <u>Phys. Rev.</u>, 92:1152 (1953).
22. A.K. Jonscher, Charging and Discharging of Non-Ideal Capacitors, IEEE Trans. Elect. Ins., in press.
23. A.K. Jonscher, Energy Losses in Charging and Discharging of Capacitors, IEEE Trans. Elect. Ins., in press.

SOLID STATE OPENING SWITCHES

Manfred Kahn

Code 6360 Ceramics Branch
Naval Research Laboratory
Washington, DC 20375

INTRODUCTION

Scope of the Problem

The generation of pulse power through the use of inductive energy storage devices frequently requires an opening switch that can operate repetitively, tolerates an instantaneous voltage increase, has a low inductance and carry a heavy load. The application of nonlinear, solid state materials to implement these types of opening switches is described in this chapter.

Approach

The magnitude of the current in a particular volume of material depends upon the carrier concentration in that volume and also on the mobility of the carriers. In an arc or plasma the current can be reduced only by lowering the carrier concentration, as for example by cooling the ionized channel by sudden contact with a wall. The current in a solid material on the other hand, can also be lowered by reducing the mobility of the carriers. This can often be done by heating the conducting material. Such a switching mode makes it possible to increase the voltage drop across the device at a rate determined by its temperature rise. There is no need for a delay as might be imposed by the time required to remove the carriers from a plasma. Consequently it appears possible, in principle, to increase the voltage across a solid faster than it can be raised in a gas or vacuum. These devices can also be switched repetitively, in contrast to single shot devices, such as fuses. In addition it is possible to control the device "on-time" by varying the preswitching load current density.

The utilization of solid state bulk switching devices has therefore been considered for the aforementioned applications.[1] Because of intrinsic energy density limitations (due to material resistivities, avalanche breakdown, thermal constraints, etc.) bulk power switching devices will, of necessity have a substantial volume. Performance optimization thus requires uniformity of energy density. For this the device should be homogeneous and also be switched homogeneously. Load sharing of multiple switching devices may have to be used. The greater the capability of the individual switches, the smaller the switching array.

SOLID STATE SWITCHING CONFIGURATIONS

Resistivity changes in a solid can stem from the transition of a superconductor to a regular conductor, from a regular conductor to a semiconductor or from a semi-conductor to an insulator. Of interest here are materials where such a change occurs reversibly, over a convenient range of an input variable, such as temperature, that triggers the transition. These materials are to be contrasted with the conversion of a fuse from the metallic to the vapor phase, where the fuse cannot be reconstituted after the transition to a high resistivity has occurred.

A nonlinear change in the resistivity of semiconducting materials is often attained by a phase transformation that causes a change in the structure of the materials. Many resulting resistivity changes are sufficiently large to be exploited in pulse power switching. As such changes are reversible, opening switches made from these materials can be recycled.

There are a number of configurations of solid state switches, that each have unique features:

1. Conventional silicon devices have planar p-n junctions that concentrate the switching losses in a thin, almost two-dimensional zone. Their application to high power switching requires use of a heat sink applied to a large area; even then the low thermal mass of the junction can easily lead to unacceptable, highly localized, temperatures during rapid switching, thereby damaging the junction permanently. Junction devices switch by virtue of changes in charge carrier concentration. "Off" switching then requires fast annihilation of carriers, either by recombination or by sweepout. Both techniques impose undesirable design compromises in large power devices. Planar silicon junction devices will not be considered further in this chapter.

2. Some low resistance polycrystalline materials can develop electron barriers at the grain boundaries. These comprise, in effect, an interconnected three-dimensional series-parallel network of switching junctions, interspersed with "inactive" carrier material. During switching, the carrier material serves as a convenient sink for the excess heat generated in the switching regions. This reduces the rate of temperature rise in the blocking barriers and moderates the temperature of any hot spots that may form. Furthermore, the shorting of a single barrier has no noticeable effect in the 3 dimensional grain boundary network, effectively providing longer life time for such devices. When the grain boundaries are switched to a high resistance, they must hold off the applied voltage since there is then no significant voltage drop across the lower resistivity grain bulk regions. Such partial utilization of the material increases the switch volume relative to that needed by a true bulk (homogeneous) switch. In addition, this extra volume tends to contribute to the resistance of the device during conduction, requiring an even larger volume to maintain a given low "on" resistance. As stated, the advantage of the added volume is that it acts as a sink for excess heat generated in the boundary layers.

3. A true bulk switching material exhibits a changeable carrier mobility throughout all of its volume. Such materials can provide, in principle, the highest volumetric efficiency, i.e., they utilize the entire volume of material uniformly. As in the above configurations, the resistivity change is triggered by an external input, such as a change in temperature, pressure or magnetic field.

BULK SWITCHING MATERIALS

The list of candidate materials extends from some presently in commercial usage to exotic materials of ill-defined preparation and purity, evaluated in minute samples at low signal levels. A cross-section of these is described below.

Barium Titanate based Grain Boundary Switches

This ceramic can exhibit a sharp positive temperature coefficient of resistance (PTCR effect) above 100°C.[2,3] Depending upon signal amplitude the resistance increase ranges from 10^3 to 10^7. This is presently being utilized in 60 Hz motor starting switches and similar applications.

These switches are polycrystalline, multiphase devices, in which a phase transition in the grains causes a conductivity change in the grain boundaries. Essentially the grain bulk acts to hold and condition the grain boundary phase. The grain bulk is also a source of current carriers when the grain boundary phase is in its low resistance condition. The division of functions facilitates the optimization of the different phases during device manufacture: The grain boundary phase responds to changes in composition and processing in a different way than the grain bulk phase. Specifically, the dissolution of a small amount (< 0.25%) of a donor dopant (i.e., Lanthanum or Niobium) makes normally insulating Barium Titanate semiconductive, so that it attains resistivities as low as 10-Ω-cm. After sintering, a lower temperature annealing treatment is applied to diffuse oxygen into the grain boundaries, creating acceptor sites. The much lower oxygen diffusivity of the donor-doped grain bulk prevents this from happening in the grain bulk. The grain boundary acceptor sites create a space charge that impedes the flow of conduction electrons, thereby raising the resistance at the grain boundaries significantly.

Below 125°C, Barium Titanate contains poled domains and charges in the domain walls. Grain boundaries are usually coincident with the walls of domains, and as a result Barium Titanate, has charged grain boundaries. Half of these contain negative charges, that neutralize the space charge from the oxygen-induced acceptor states and permit easy flow of conduction electrons across the grain boundaries. At the 125°C transition temperature of $BaTiO_3$, the domain walls and their charges disappear.[4] The grain boundary resistance goes up and the effective resistivity of the ceramic increases by three or more orders of magnitude. This is then the noted PTCR effect.

Carbon Filled Polymer (CFP) PTCR Devices

These devices are presently used to protect low voltage (< 50 V) circuits. They are polymer-carbon composites that increase their resistance, upon heating, from two to five orders of magnitude.[5] They are made of polyethelene that is filled with about 50% of carbon particles. The carbon links into chains that lead to relatively low resistivity (about 1 Ω-cm) at ambient temperature. Heating of the device above 85°C causes a phase change in the polymer that is accompanied by a volume increase of about 2%. This breaks the conducting carbon chain. Thin insulating regions then form between the carbon particles, increasing the resistance of the device significantly. Upon cooling, recrystallization of the polymer and re-linking of the chains occur and the device recovers to approximately 90% of its initial conductivity. Test devices have been cycled tens of times without further degradation. When left inoperative, re-linking of carbon chains continues, with recovery to essentially the initial state over a 24 hour period.

Carbon filled polymer is normally prepared in thin sheet form. This provides devices that can have 1/10 to 1/100 of the resistance of $BaTiO_3$ based PTCRs. At the same time this makes higher voltage designs more difficult to implement. These devices exhibit a resistance drop when external pressure is applied, since this brings the carbon particles into more intimate contact.

Indium Antimonide

This material exhibits a bulk resistivity increase when immersed in a magnetic field. This is a result of its extraordinarily high carrier mobility: a transverse magnetic field then causes the current to be deflected away from the opposite electrode. This has been evaluated in an experimental power switch,[6] where this property is utilized in a doughnut shaped indium antimonide "Corbino" disc,[7] inserted into a strong magnetic field of about 5 Tesla. The current carriers then follow a spiral path, increasing the effective resistance by as much as two orders of magnitude. Current densities can be as high as 30 A/mm^2 and a voltage withstand capability of 1.5 kV/cm can be attained. This material appears to be a candidate for a high power switch in systems that generate large magnetic fields. It can be switched at room temperature and in its polycrystalline form no significant grain boundary resistance has been observed. It melts at 535°C, crystals are relatively easy to grow and Czochralski pulled boules in excess of 5 cm in diameter can be fabricated. Casting of polycrystalline ingots should permit the preparation of even larger pieces.

Europium Oxide

Single crystal samples have been shown to exhibit a bulk metallic-to-semiconductor transition at about 50°K. Their resistivity (typically 0.005 Ω-cm) increases by as much as 5 orders magnitude with rising temperature.[8] In addition, the resistivity transition is shifted to higher temperatures in the presence of a magnetic field. This property is particularly interesting, since it can be used to maintain a uniform transition in a large volume switch by immersing it in the magnetic field of the current controlled by the switch. The transition uniformity is attained through the resulting feedback loop, where the onset of switching in one segment of the switch, leads to a reduction of the current, lowers the field and induces switching in other parts of the material which has remained slightly cooler.

Chromium Doped Vanadium Sesquioxide

This material develops a bulk PTCR effect[9] at about 80°C. Small single crystals increase in resistance by about two orders of magnitude, larger experimental polycrystalline devices show a significantly smaller change. In this material the presence of grain boundaries appears to dilute the PTCR effect of the bulk of the grains.

Superconducting Materials

The "normal" resistivity in these is attained as a result of heating or of an increase in the magnetic field (e.g., increasing the current above the critical limit). The abrupt transition from zero to a finite resistance has been applied in the development of experimental superconducting switches.[10] Their main limitations are the relatively low operating voltage and the low resistivity of the switches in their "off" state. A high temperature (normal) resistivity of up to 0.1 Ω-cm seems attainable in some materials. In addition it becomes quite cumbersome to control the heat flow through the lead-in wires.

SWITCHING MODES AND CONSTRAINTS

PTCR devices, i.e., as made from materials as described previously, can be operated as opening switches by increasing their temperature (and resistance) through Joule heating. This can be applied at a high rate, limited only by thermal shock from nonuniform temperature increases, as discussed later.

There are a number of materials that exhibit a sharp drop in resistivity with increased temperature. They require cooling by conduction of heat to open a circuit. This is a relatively slow process and is not considered here. On the other hand, a magnetic field to increase the resistivity of a magneto-resistive material can be established within microseconds and the application regime of such a material bears further evaluation.

Operation of PTCR Devices

The small signal resistance-temperature characteristics of $BaTiO_3$ and of carbon filled polymer are shown in Fig. 1. It shows two operating regions with greatly different temperature-dependent resistance behavior: The first region below temperature $T_d \sim 100°C$, has a relatively linear temperature dependence of resistivity, with even a slightly negative slope in $BaTiO_3$. The delay of switching on-set t_d, is inversely proportional to the power input, (i.e., $P = I^2R$), that heats the PTCR switch up to T_d:

$$t_d = (M \int_{T_o}^{T_d} C_v \, dt)/P \qquad (1)$$

Here T_o is the ambient temperature, M is the mass of the switching material, and C_v its specific heat capacity (0.5 J/g °C, for

SOLID STATE OPENING SWITCHES

BaTiO$_3$, for instance). Assuming that in pulsed operation, no significant heat is lost to the environment by the switch, the delay time is

$$t_d = M C_v (T_d - T_o)/P \qquad (2)$$

For a 100°C ceramic PTCR switch which is initially at $T_o = 25°C$, the delay is

$$t_d = 37.5 \frac{\text{joules}}{\text{gram}} \times \frac{\text{mass(gr)}}{\text{power(W)}} \qquad (3)$$

As an example, a BaTiO$_3$ sample weighing 20 grams absorbs before switching a power density of 0.75 kW/g, giving a delay time of 0.5 sec.

To determine the switching interval itself, the procedure outlined in Ref. 11 can be used. The assumption of zero heat loss during switching leads to a simple relation, that shows that the conducting material increases in temperature, T, at a rate proportional to the specific power input,

$$\frac{dT}{dt} = \frac{\rho j^2}{C_v}, \qquad (4)$$

where C_v is again the specific heat capacity, ρ is the resistivity of the material and j is current density. Approximating the resistance characteristics shown in Fig. 1 above T_d by an exponential, i.e., $\rho(T) = \rho_o e^{\alpha(T-T_d)}$ for $T>T_d$, the solution of Eq. (4) becomes

$$e^{-\alpha(T-T_d)} = 1 - \frac{\alpha \rho_o j^2 t}{C_v} \qquad (5)$$

where t is the current fall time interval. If the initial resistance increases by γ, such that $\rho = \gamma \rho_o$, then

$$\frac{1}{\gamma} = 1 - \frac{\alpha \rho_o j^2 t}{C_v} \qquad (6)$$

and

$$t = C_v / \left(\alpha \rho_o j^2 \right), \text{ for } \gamma \gg 1. \qquad (7)$$

Equation (7) can be written in terms of the applied field, $E = \gamma j \rho_o$,

$$t = C_v \rho_o \gamma^2 / E^2 . \qquad (8)$$

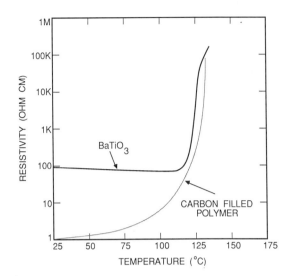

Fig. 1. PTCR characteristics of barium titanate and of carbon-filled polymer.

One limiting factor in Eq. (8), is the maximum hold-off field of the switch, E_m. To obtain a minimum current fall time, t_{min}, the maximum value of j that can be passed by the non-linear material is used (i.e., $j = E_m/\gamma\rho_o$). The value of t_{min} then is:

$$t_{min} = C_v \rho_o \gamma^2 / E_m^2 \qquad (9)$$

This relationship indicates the importance of the breakdown voltage of the PTCR in its ability to switch rapidly. Additionally, other effects, such as thermal shock may limit the performance further. After switching, the off condition of the switch is maintained by current leakage. This adjusts itself for the I^2R dissipation to equal the heat loss of the PTCR, as long as the temperature is maintained within the range where the resistance of the material increases with temperature.

Limitations of PTCR Switches

(1) Figure 2 shows schematically how the resistance-temperature behavior of barrier type PTCR devices depends upon voltage amplitude. The resistivity may decrease by 50% in a cold device, but above the switch temperature, where the voltage appears only across the relatively thin barriers, it can drop by two orders of magnitude.

SOLID STATE OPENING SWITCHES

BaTiO$_3$, for instance). Assuming that in pulsed operation, no significant heat is lost to the environment by the switch, the delay time is

$$t_d = M C_v (T_d - T_o)/P \tag{2}$$

For a 100°C ceramic PTCR switch which is initially at $T_o = 25°C$, the delay is

$$t_d = 37.5 \frac{\text{joules}}{\text{gram}} \times \frac{\text{mass(gr)}}{\text{power(W)}} \tag{3}$$

As an example, a BaTiO$_3$ sample weighing 20 grams absorbs before switching a power density of 0.75 kW/g, giving a delay time of 0.5 sec.

To determine the switching interval itself, the procedure outlined in Ref. 11 can be used. The assumption of zero heat loss during switching leads to a simple relation, that shows that the conducting material increases in temperature, T, at a rate proportional to the specific power input,

$$\frac{dT}{dt} = \frac{\rho j^2}{C_v}, \tag{4}$$

where C_v is again the specific heat capacity, ρ is the resistivity of the material and j is current density. Approximating the resistance characteristics shown in Fig. 1 above T_d by an exponential, i.e., $\rho(T) = \rho_o e^{\alpha(T-T_d)}$ for $T>T_d$, the solution of Eq. (4) becomes

$$e^{-\alpha(T-T_d)} = 1 - \frac{\alpha \rho_o j^2 t}{C_v} \tag{5}$$

where t is the current fall time interval. If the initial resistance increases by γ, such that $\rho = \gamma \rho_o$, then

$$\frac{1}{\gamma} = 1 - \frac{\alpha \rho_o j^2 t}{C_v} \tag{6}$$

and

$$t = C_v / \left(\alpha \rho_o j^2 \right), \text{ for } \gamma \gg 1. \tag{7}$$

Equation (7) can be written in terms of the applied field, $E = \gamma j \rho_o$,

$$t = C_v \rho_o \gamma^2 / E^2 . \tag{8}$$

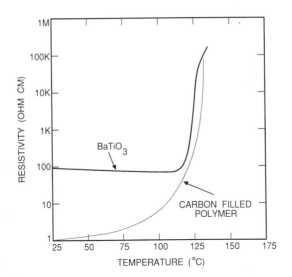

Fig. 1. PTCR characteristics of barium titanate and of carbon-filled polymer.

One limiting factor in Eq. (8), is the maximum hold-off field of the switch, E_m. To obtain a minimum current fall time, t_{min}, the maximum value of j that can be passed by the non-linear material is used (i.e., $j = E_m/\gamma\rho_o$). The value of t_{min} then is:

$$t_{min} = C_v \rho_o \gamma^2/E_m^2 \qquad (9)$$

This relationship indicates the importance of the breakdown voltage of the PTCR in its ability to switch rapidly. Additionally, other effects, such as thermal shock may limit the performance further. After switching, the off condition of the switch is maintained by current leakage. This adjusts itself for the I^2R dissipation to equal the heat loss of the PTCR, as long as the temperature is maintained within the range where the resistance of the material increases with temperature.

Limitations of PTCR Switches

(1) Figure 2 shows schematically how the resistance-temperature behavior of barrier type PTCR devices depends upon voltage amplitude. The resistivity may decrease by 50% in a cold device, but above the switch temperature, where the voltage appears only across the relatively thin barriers, it can drop by two orders of magnitude.

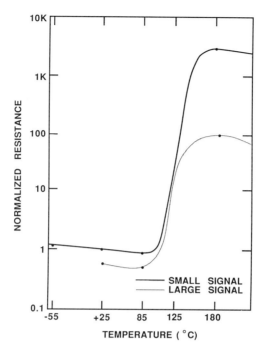

Fig. 2. 125°C PTCR resistance-temperature characteristic of a barrier type device as a function of voltage amplitude.

(2) There is a limiting temperature, above which the resistance of a PTCR does not increase with additional heating. A PTCR heated into this temperature range by a relatively low impedance supply i.e., by a voltage source, will experience a resistance drop, thermal runaway and potential destruction. In pulse testing, on the other hand, the high voltage is applied only for a short period. The heat capacity of the device then tends to prevent thermal runaway, permitting higher than "steady state" voltages. It is possible, nevertheless, that thermal runaway can still appear under pulsed conditions since resistivity inhomogeneities exist in just about all real devices. Figure 3a shows cracks that can develop in a ceramic PTCR with a lower resistivity sector that has gone into thermal runaway. Heat conduction then can become too slow to prevent the excess current density in this sector from introducing a damaging temperature gradient and destructive expansion stresses.

There are also intrinsic thermal gradients in PTCR's.[12] These occur because the electrodes comprise a thermal mass that retards heating near them just long enough to allow the center of the device to increase in resistivity before the outer regions do.

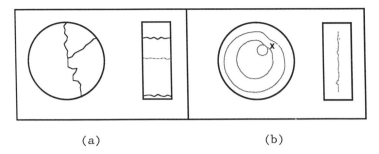

Fig. 3. (a) Axial cracks from excessive steady state voltages in a non-homogeneous PTCR.
(b) Radial delamination from excessive heat-up power in a homogeneous PTCR.

This is shown in Fig. 4. The higher resistivity regions are subjected to increased power dissipation densities, that can again lead to excessive thermal stresses and to radial cracking, as shown in Fig. 3b. The uneven resistivity distribution also causes a significant field enhancement near the resistivity peak and air ionization has been observed at the instant of failure. The magnitude of these effects depends on device structure and packaging as much as on material parameters. It becomes quite evident that an

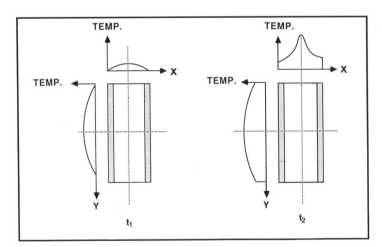

Fig. 4. Temperature distribution in a discoidal PTCR during fast heat-up. Y is the radial direction and X is the thickness, t_1 is the instant before the center of the part attains 85°C. At t_2 the resistivity increase in the central Y plane causes most of the power dissipation to be concentrated there, with an accompanying cumulative localized temperature rise.

SOLID STATE OPENING SWITCHES

increase in the thickness of a PTCR will result in a proportional increase in voltage withstand capability only under steady state conditions.

Magnetoresistive Switching

As indicated, a high magnetic field and possibly cooling to below normal ambient is needed to switch indium antimony effectivity.

Work with polycrystalline[6] InSb highlighted some of its power density and efficiency limitations. Improved quality and the use of single crystals as well as a means to utilize the field generated by the storage coil may well result in practical device. As shown in Fig. 5, its resistance can increase by a factor of 50 in a magnetic field of 5 Tesla, with a potential switching speed of <10 µs.

PULSE POWER TEST RESULTS

It can be expected that switching materials will behave differently when subjected to high amplitude, short pulses, than when smaller signal, continuous power is applied.

Specifically, tests on 20 volt rated $BaTiO_3$ PTCR devices (Keystone Carbon RL5405-30-120-20-PTO), 1.4 cm diameter and 0.12 cm thick, indicated that their steady state runaway voltage was about

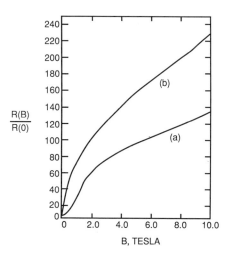

Fig. 5. Resistance vs applied magnetic field, (a) room temperature, and (b) 77°K, for InSb.

60 V. Repeated applications of 400 V pulses for times in excess of 1 ms were tolerated without damage to the device.[13] "Normal" operation of the device is with 6 A, which causes it to switch after 0.4 seconds. In low impedance pulse testing, currents in excess of 40 A were applied. These caused the start of switching after 80 ms and then the superposition of a judiciously timed trigger pulse gave a current fall time of 1 ms.

Tests were also conducted on carbon filled polymer PTCR devices (Raychem PSR 20528), that have conventional voltage and current ratings of 15 V and 9 A. They also had thermal runaway below 75 V, but they tolerated 10 kA pulses and 175 V spikes for 1 ms periods without damage.

The high voltage, hot resistance of both types of devices was not more than 100 times their cold resistance, a significant reduction from the ratios observed with small signals. This ratio nevertheless has been considered adequate for at least some pulsed power switching applications.

Carbon filled polymer PTCR's have also been evaluated as disposable, one shot devices, i.e., as a direct substitute for a fuse. A high impedance pulse source applied 2000 A to the mentioned 15 V, 9 A parts, which caused them to interrupt after 2.5 ms with a current fall time of 100 μs. The parts then self destructed, allowing the power supply voltage to rise to 600 V before restrike occurred. Their performance in this respect was at least as good and more consistent than that of 5 cm long wire fuses designed for a similar switching time. This would imply that after the voltage induced destruction of a carbon filled polymer with silver electrodes, the restrike voltage stress is more than an order of magnitude larger than that across a copper wire fuse blown up in air.

The voltages and currents measured on these PTCR devices under pulsed power conditions are significantly higher than those deduced from steady state testing. It nevertheless is still quite obvious that arrays of devices will have to be used to meet system requirements: Testing of PTCR's in parallel shows equitable current sharing and rapid and near syncronous shutoff, triggered by the fastest device. Increased voltage withstand capability on the other hand can be obtained by a series connection only where voltage sharing is ensured, i.e., overvoltage protection devices are applied across each series PTCR. Resistors, capacitors or varistors can be used for this.

ACKNOWLEDGMENT

The assistance of I. Vitkovitsky and R. Ford in the compilation of this material is hereby gratefully acknowledged.

REFERENCES

1. R.D. Ford, I. Vitkovitsky, and M. Kahn, Application of Nonlinear Resistors to Inductive Switching, *IEEE Trans. Electrical Insulation*, 20: 29 (1985).
2. W. Heywang, Resistivity Anomaly in Doped Barium Titanate, *So. Am. Cer. Soc.* 47: 484 (1964).
3. M. Kahn, Effects of Heat Treatments on the PTCR Anomalie in Barium Titanate, *Bulletin of the American Ceramic Society*, 50: 676 (1971).
4. G.A. Jonker, Some Aspects of Semiconducting Barium Titanate, *Sol. State Electr.* 7: 895 (1964).
5. A.F. Doljak, Device for Overcurrent Protection, *IEEE Trans. on Components, Hybrids, and Manufacturing Technology*, CHMT-4 :372 (1981).
6. E.K. Inall, A.E. Robson, and P.J. Turchi, Application of the Hall Effect to the Switching of Indoctive Circuits, *Rev. Sci. Inst.*, 48: 462 (1977).
7. D.A. Kleinman and A.L. Schawlow, Corbino Disk, *Journal of Appl. Phys.* 31: 2176 (1960).
8. Y. Shapiro and S. Foner, Resistivity and Hall Effects in Fields up to 150 KOe, *Phys. Rev. B.*, 8: 2299 (1973).
9. R.S. Perkins, A. Ruegg, M. Fischer, P. Streit, and A. Menth, A New PCT Resistor for Power Applications, *IEEE Trans. on Components, Hybrids Manufacturing and Technology*, CHMT-5: 225 (1982).
10. V.A. Glukhikh, A.I. Kostenko, N.A. Monoszon, V.A. Tishchenko and G.V. Trokhachev, Results of Investigation of High Speed Breaking Superconducting Switch, Proc. 7th Sym. on Engineering Problems of Fusion Research, 1: 912, Knoxville, TN (1977).
11. R. Ford and I.M. Vitkovitsky, Inductive Storage Pulse Train Generator, *IEEE Trans. on Electron Devices*, ED-26: 1527 (1979).
12. M. Kahn, Application Principles and Limitations of PTC Resistors, *Ferroelectrics*, 11: 331 (1976).
13. R. Ford, Personal Communication, NRL, Code 4775, Wash., D.C. (1984).

MECHANICAL SWITCHES

W. M. Parsons

Los Alamos National Laboratory
Los Alamos, NM 87545

INTRODUCTION

Since the 1700's when scientists first began conducting experiments with electrical conductors, mechanical switches have been utilized to control the flow of electrical current. At first these switches consisted of no more than a pair of either separated or connected wires. Simple mechanical switches, similar to knife switches, were soon developed to facilitate the making and breaking of small electrical currents. The electrificiation of cities and industry in the early 1900's led to the development of mechanical switches capable of carrying and interrupting much higher currents and voltages. The first power circuit breakers were built in the 1920's and used air as the arc extinguishing medium. These devices could interrupt several kiloamperes at voltages of ~ 15 kV. Since that time, extensive research and development has led to switches capable of carrying and interrupting even much higher currents and voltages. Modern SF_6 filled circuit breakers are able to operate at system voltages as high as 765 kV. Experimental switches for electromagnetic launchers are being designed for operation at currents in excess of 1 MA.

The purpose of this chapter is to analyze the existing technology in high power mechanical switches and to comment on their suitability for repetitive operation. Closing and opening switches will be treated in separate sections. Several experimental switches will be discussed, but primary emphasis will be placed on commercial switches and modified commercial switches. The basic theory of operation, current and voltage ratings, opening and closing speeds, and some weight-volume considerations will be included for each generic type of switch. Finally, switching for repetitive operation will be discussed in terms of potentially achievable repetition rates.

OPERATING MECHANISMS

By definition, mechanical switches are devices that rely on the physical movement of contacts or insulators to provide electrical switching. These devices frequently employ linkages, levers, camshafts, rotating drums or discs, and other mechanical parts for actuation. The power required for insulator or contact acceleration is usually supplied by energy storage systems such as springs, compressed air, or electrical capacitors.

Most mechanical switches are designed primarily for either opening or closing and generally utilize the energy storage system for accomplishing that particular operation. The reverse operation generally takes place on a much slower time scale. In either case, a series of valves must be opened, latches tripped, or springs compressed for the switch to change state and then reset with limits imposed on their speed of operation, e.g., pneumatic switches are limited by gas sonic velocities (~ 300 m/s) through charging or relief valves. This series of events ultimately limits the achievable repetition rate for high-power mechanical switches. Most standard commercial switchgear mechanisms are designed to operate in the 0.5 to 2 Hz regime with no modification. Specially designed mechanisms could probably achieve operation at several tens of hertz. However, any mechanism is limited to the wave propagation rate through the linkages. Higher voltages require longer linkages thus increasing actuation time. Higher currents require more massive contacts and linkages increasing their actuation time from inertial limitations.

CLOSING SWITCHES

Closing switches are switches that are require to "make" an electrical connection and carry current, generally under some open circuit voltage. The duties of closing switches are generally less severe than those of opening switches, and their design is easier. The two major design criteria for mechanical closing switches are:

1. Sufficient speed during closing to avoid excessive arcing following prestrike.

2. Bounceless closing ("making") operation to avoid arcing during contact closure.

MECHANICAL SWITCHES

Adequate closing speed is necessary because any switch will prestrike when the contact gap becomes too small to withstand the impressed open circuit voltage. The switch will conduct by arcing from the time of prestrike until galvanic closure is achieved. Since arcing causes contact erosion, this time is minimized by bringing the contacts together at a high rate of speed. Similarly, contact bounce after closure creates arcing which deteriorates contact surfaces, and, in some cases, may actually weld the contacts together. Unfortunately, rapid closing of contacts is the major cause of contact bounce, so the designer is faced with a trade-off between the two requirements.

Several varieties of high power closing switches are currently available on the commercial market. These are generally low voltage (<600 V) switches with high continuous current ratings and are used extensively in the electroplating and aluminum refining industries. Several of these switches can be easily modified for increased voltage or current ratings. One such modification will be discussed in detail in the next section.

Many varieties of experimental closing switches have been developed in recent years for specialized applications. Of special interest are the metal to metal, magnetically driven, closing switches with solid or gaseous insulation. Several of these will be described later.

COMMERCIAL CLOSING SWITCHES

Two manufacturers of commerical high power closing switches are the Pringle Company and Brown Boveri Electric, Inc., formerly ITE. Pringle uses a knife-switch design whereas Brown Boveri uses a rocker contact concept. Both type switches avoid contact bounce through innovative design. The Pringle switch has a massive armature, which, for the larger models (designed for tens of kiloamperes), operates from a motor driven worm gear. This switch takes several seconds for opening and closing and is not particularly suitable for fast repetitive operation. A photograph of a typical Pringle switch is shown in Fig. 1. The rocker contact switch manufactured by Brown Boveri Electric, Inc. moves a very small bridge contact between two stationary electrodes. A simplified schematic of this switch is shown in Fig. 2. Typical ratings for this type of switch is 9 kA at 600 V. This switch[1] was immersed in silicone oil during an experiment and run at voltages as high as 60 kV. Further modifications to increase the current carrying capacity resulted in a device that was capable of carrying 25 kA on a continuous basis[2]. A photograph of this switch in a switch testing facility is shown in Fig. 3. Opening and closing times for these modified switches using commercial 125 psi air cylinders as

Fig. 1. Pringle closing switch.

actuators are typically on the order of 40 ms. Repetition rates of 10 Hz would be quite possible.

A wide variety of utility disconnect switches, such as those manufactured by S & C Electric Company, can be used as closing (make) switches. These switches[3] usually operate within 5 to 8 cycles of 60 Hz (i.e. 5/60 to 8/60 sec), have steady-state current ratings of 600 to 2000 A, momentary current ratings as high as

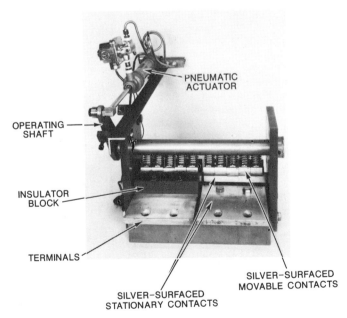

Fig. 2. Brown Boveri rocker contact closing switch.

MECHANICAL SWITCHES

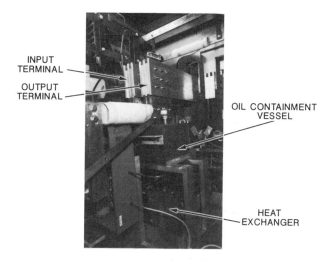

Fig. 3. Modified rocker switch.

80 kA, a fault closing duty of 40 kA (i.e. closing on 40 kA rms current) as high as 500 kV. They are generally quite large devices designed for outdoor use and are normally mounted on distribution line towers or in switchyards. Detailed information concerning this category of mechanical switches can be obtained from the manufacturers.

EXPERIMENTAL CLOSING SWITCHES

Many varieties of experimental mechanical closing switches have been devised for specialized applications. These can roughly be categorized in two classes, low speed and high speed switches. The low speed category includes most switches with pneumatic, hydraulic, or ferromagnetic electric actuators. These usually have closing times of several tens of milliseconds. The high speed switches commonly operate on a magnetic repulsion principle. These switches close in the tens of microseconds regime and have, relatively speaking, low jitter. Some examples of both types of switches are given below.

Low Speed Devices. One example of a pneumatically driven experimental closing switch is one developed for the homopolar generator at the University of Texas at Austin[4]. This switch was designed to close on a current of 300 kA for 3 sec. The open circuit system voltage was 50 V. A schematic of the switch is shown in Fig. 4. In this switch, the pressure cylinder, 5, pulls the movable contact, 4, between the stationary contacts, 3, to make

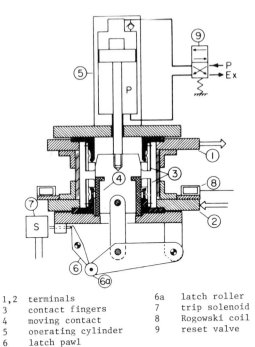

1,2	terminals	6a	latch roller
3	contact fingers	7	trip solenoid
4	moving contact	8	Rogowski coil
5	operating cylinder	9	reset valve
6	latch pawl		

Fig. 4. Pneumatically driven closing (make) switch (drawing courtesy of Paul Wildi, The University of Texas at Austin).

the circuit. This switch has been successfully tested to currents as high as 550 kA and begins to make contact 11 ms after the trip signal is given. A trip latch allows a very low dead time of 6.5 ms before the onset of motion. With some modification this switch could be used for repetitive operation in the tens of Hertz range.

High Speed Devices. The high speed switches described in this section are not particularly suitable for repetitive operation in their present form. This is due to the need for replacement of elements after each operation. If a method were devised to load the replaceable element rapidly, then repetitive operation could be possible.

One switch utilizing solid insulation with an electro-repulsive driver[5] is shown in Fig. 5. Here, the magnetic coil repels the aluminum disc, which punctures the mylar insulation, and makes contact between the stationary terminals. The authors report a system delay time of 13 µs and a jitter of 0.15 µs. The switch has been used to close on currents on high as 1.3 MA at 10 kV.

MECHANICAL SWITCHES

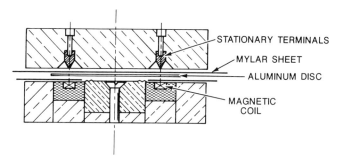

Fig. 5. Electro-repulsive switch with solid insulation.

Another switch, using air as insulation[6], is depicted in Fig. 6. In this switch, the coil, 5, repels the aluminum ring, 4, into a pair of coaxial contacts (not shown). This author tested the switch to peak currents of 121 kA at 1.1 kV. The delay time was found to be 33 μs with less than one μs jitter.

OPENING SWITCHES

As in the previous section, opening switches will be divided into two categories, commercial and experimental. The commercial switches treated will be air break, oil, air and SF_6 gas blast, and vacuum. The more novel switches involve recent developments by commercial manufacturers and some experimental switches. A special section will be devoted to the interruption of dc currents.

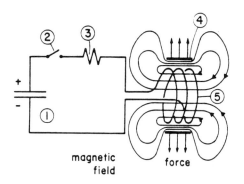

1 Capacitor
2 Switch
3 Resistance of circuit
4 Repelled load (ring, moving contact)
5 Drive coil

Fig. 6. Coaxial magnetically driven switch (drawing courtesy of Paul Wildi, The University of Texas at Austin).

Commercial Opening Switches

The development of ac circuit breakers became a scientific endeavor in the mid 1920's. The construction of short circuit test laboratories and the development of the cathode ray oscilloscope enabled scientists to study detailed phenomena associated with the breaking of electrical currents. Early ac circuit breakers required 10 to 20 current cycles to interrupt, whereas some modern circuit breakers can interrupt in as little as 1/2 cycle. The following text details the present day technology developed after many years of research on various types of mechanical circuit breakers.

Air Break Switches. The earliest interrupters were simple air break switches similar to modern knife switches. These switches had no arc enhancement or control devices and were very limited in their interrupting ability. In 1929, Westinghouse developed the Deion circuit breaker which used a metallic "splitter" to break a long single arc into a multiple series of short arcs. The added cathode drop at the foot of each of these short arcs greatly improved the dc as well as ac interrupting ability. A similar technique used a magnetic field to stretch the arc between a number of insulating fins and thereby increased the arc drop by lengthening the arc. Today, both methods are still being used by circuit breaker manufacturers. The ITE Imperial division of Brown Boveri Electric, Inc. uses the metal plate splitter while Westinghouse uses the insulating fin technique. A picture of two modern Westinghouse type DMD dc circuit breakers is shown in Fig. 7. These circuit breakers[7] are rated at 1.5 kV, 4 kA service and will interrupt dc currents as high as 80 kA. Table 1 lists the maximum

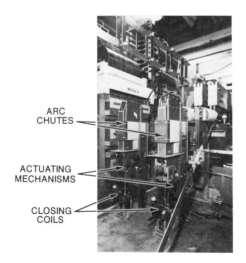

Fig. 7. Westinghouse type DMD dc circuit breakers.

TABLE 1
MAXIMUM CONTINUOUS CURRENT AND VOLTAGE RATINGS FOR VARIOUS DC CIRCUIT BREAKERS

MANUFACTURER	MODEL SERIES	MAXIMUM CONTINUOUS CURRENT (kA)	MAXIMUM RATED VOLTAGE (V)
WESTINGHOUSE[7]	DM	8	3000
	DB	6	250
	DR	10	750
	DMD	4	3000
GENERAL ELECTRIC[8]	MC	12	1000
BROWN BOVERI[9,10]	FBK	10	800
	UR 26	2.6	4000

continuous current and voltage ratings found in a recent search of manufacturer's literature for dc air circuit breakers.

The interruption rating for these breakers is much higher than the continuous rating. For instance, the GE MC6 breaker will interrupt currents[11] as high as 220 kA. Unfortunately, all interrupters of this type are inherently limited in voltage rating to a few kilovolts, although similar ac airbreak switches have ratings as high as 15 kV.

Oil Switches. The trend towards higher system voltages in the early part of the century pushed the circuit breaker designer to find an interrupting medium with better insulating properties than air. Oil, with a dielectric strength of 120 kV/cm was a logical choice. Until 1925, these circuit breakers simply involved parting a set of contacts submerged in oil. As the design became more sophisticated, use was made of the gas generated by the arcing process. This gas is used by an arc control device (an internal element of the circuit breaker) to sweep, compress, and cool the arc plasma. There are two basic types of arc control devices, axial blast and cross blast. Figure 8 illustrates how the cross blast device operates. The pressure generated by the gas (mostly hydrogen), from an arc submerged in oil, forces the plasma into a

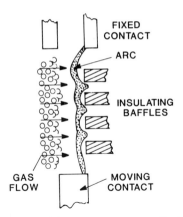

Fig. 8. Cross blast arc control device.

series of insulating baffles. This arrangement sweeps ionized by-products into the surrounding oil and cools the arc and elongates the arcing path, resulting in a higher arc voltage. When a current zero occurs, the interrupter is less likely to restrike on the resulting recovery voltage. The axial blast arc control device works in much the same way except that the insulating baffles are arranged in an annular fashion about the contacts.

Oil circuit breakers are generally classified as the live tank or dead tank type. In the live tank variety, the oil is usually contained by a tank which is insulated from ground. These circuit breakers contain a minimal amount of oil and, though often quite tall, are not extremely bulky. Dead tank circuit breakers, on the other hand, contain a huge volume of oil and are quite bulky. The oil is held in a steel container which is normally grounded. These switches can weigh many tens of tons for the high voltage level ones.

Typical closing (making) and opening (breaking) speeds[12] of oil circuit breakers are listed in Table 2. Typical interruption ratings[13] for oil circuit breakers are listed in Table 3. The ratings listed are for symmetrical fault currents.

Air Blast. The first air blast circuit breaker was developed by Whitney and Wedmore[14]. The general operating principle behind these circuit breakers is the use of compressed air flow to cool and increase the resistance of an electrical arc between contacts. In fact. these interrupters are so adept at increasing the rise of resistance near current zero that parallel resistors must sometimes be added to prevent over-voltages due to current chopping (i.e. L di/dt voltages).

TABLE 2

TYPICAL MAKING AND BREAKING SPEEDS OF OIL CIRCUIT BREAKER CONTACTS

VOLTAGE CLASS (kV)	BREAKING SPEED (m/s)	MAKING SPEED (m/s)
11	2.4	3.3
33	4.3	5.2
132	7.6	8.5

TABLE 3

SYMMETRICAL INTERRUPTION RATINGS FOR OIL CIRCUIT BREAKERS

VOLTAGE CLASS (kV)	INTERRUPTION RATINGS (kA)	INTERRUPTION TIME (CYCLES)
15.5	18	5
38	22	5
72	26	5
121	63	3
242	63	3

The two basic categories of air blast circuit breakers refer to the arc control device which is utilized. These two categories are axial blast and cross blast. The cross blast design is used only in low pressure auxiliary switches due to its inability to produce high air velocities in the arcing region. The axial blast arc control is used exclusively in modern circuit breakers and can be a mono-blast or a duo-blast type. The mono-blast simply forces the air axially along the entire arc. In the duo-blast system, the pressurized incoming air is divided into two streams, each of which flows along half of the arc but in opposite directions.

TABLE 4

SYMMETRICAL INTERRUPTION RATINGS FOR AIR BLAST CIRCUIT BREAKERS

VOLTAGE INTERRUPTION CLASS (kV)	INTERRUPTION RATINGS (kA)	TIME (CYCLES)
24	63	2
550	63	2
800	63	2

Air blast circuit breakers are very popular due to their inherent safety since they utilize neither flammable liquids nor toxic gases. Their ability to break heavy fault currents together with their design flexibility allows voltage capabilities to be altered by changing the design pressure, and their voltage rating per interrupter is high. As mentioned previously, current chopping at low currents sometimes requires ancillary components. These circuit breakers are, however, quite limited in their ability to withstand a high rate of rise of recovery voltage.

Typical interruption ratings[13] for air blast circuit breakers are listed in Table 4. The ratings listed are for symmetrical fault currents. Air blast circuit breakers are acoustically very noisy when interrupting but can be used indoors with silencers which limit the noise level to 105 db.

SF_6 Gas Blast. SF_6 gas was first applied as an insulating medium in the early 1940's. It has several unique properties which make it attractive for use as an arc quenching medium in circuit breakers. They are:

1. SF_6 has a high insulating level compared to air. Fifteen atm of SF_6 equals 50 atm of air in insulating capability.

2. It recovers dielectric strength very quickly after passing an arc.

3. It has a greater thermal conductivity than air.

4. It has no carbon atoms in its molecular composition to contaminate insulating surface within the interrupter.

5. The lack of oxygen in the system prevents oxidation of electrical contact surfaces.

There are two types of commercial SF_6 blast circuit breakers, the two-pressure system and the puffer. The two-pressure system is very similar to the air blast breaker except that it is a closed system. The puffer uses the piston motion of the actuating mechanism to force gas through the arc.

The first domestic two-pressure system was developed by Westinghouse in the late 1950's. As with the air blast device, parallel resistors were sometimes added to prevent over-voltages due to the high rate of change of arc resistance during the interruption process. Also in the 1950's, a puffer system was developed by Westinghouse. This type of interrupter has gained increased popularity due to the simplicity of the system. Today, modular SF_6 puffer interrupters are assembled in series for use at voltages as high as 765 kV. They are also being considered for use at the new utility industry EHV levels of 1050 kV.

Vacuum Circuit Breakers. Switching experiments in vacuum began at the California Institute of Technology in 1923. Sorensen and Mendenhall[15] reported interrupting 900 A at 40 kV. However, due to problems in the glass to metal seals and the outgassing of materials, vacuum switches did not become practical until the 1950's. In 1962, General Electric announced the development of the first vacuum circuit breaker that could interrupt fault-like currents. The breaker was rated for 12.5 kA at 15.5 kV. Recent developments[16] in this technology have produced interrupters which have ac symmetrical interruption ratings as high as 100 kA. Some types are especially manufactured for high voltages, such as the GE PV08. This interrupter is rated[17] at 45 kV and can interrupt symmetrical currents of 40 kA. The popularity of the vacuum interrupter can be attributed to several reasons:

1. The devices are self-contained, quiet, and very simple to operate. Their relatively short stroke (6-8 mm) allows a compact design, a low actuating energy, and thus an operating mechanism that can be made quite reliable.

2. They have the fastest dielectric recovery time of any commercial power circuit breaker. Lee[18] reports dv/dt withstand after interruption of up to 24 kV/μs. Indeed, many manufacturers specify interruption times of 1/2 to 1 cycle.

3. They are inexpensive. The present day vacuum interrupter replacement tube cost is about $1500.

The vacuum interrupter has also been used extensively for dc breaking applications in magnetic fusion experiments. A photograph of three, single-phase vacuum interrupters with actuators and drivers is shown in Fig. 9.

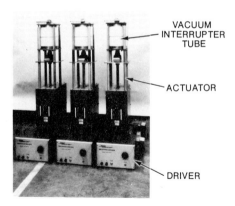

Fig. 9. Single phase vacuum interrupters with actuators and drivers.

Recent Developments. One of the newest innovations in interrupter technology was made by Brown Boveri Electric, Inc. This was the SF_6 arc-spinner interrupter. In this design an electric arc is moved through stationary SF_6 at about 5 atm. The arc is rotated on a set of ring contacts by means of a self-generated magnetic field. The advantages possessed by this type of interrupter are:

1. An exceptionally high recovery speed allowing rate of rise of recovery voltages of several kV per microsecond.

2. A high dielectric strength inherent with SF_6, permitting operation at high voltages.

3. Mechanical and electrical simplicity, which minimizes capital investment and insures a high degree of mechanical reliability.

4. Low contact erosion due to the fact that the arc is always moving and cannot become rooted in one spot.

Another recent innovation was made by the Los Alamos National Laboratory and Westinghouse Electric Corporation. This was the development[19] of a water-cooled vacuum interrupter capable of steady-state current conduction or interruption at 25 kA. A picture of this interrupter in a Los Alamos switch facility is shown in Fig. 10. This switch has added potential for repetitive operation due to the heat removal system designed for high, steady state conduction.

A mechanical switch is presently being developed for use in repetitively operated electromagnetic launcher systems by General

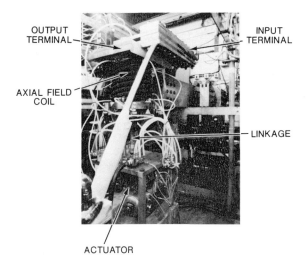

Fig. 10. Water-cooled vacuum interrupter.

Dynamics and IAP Research. The concept was first tested by Barber and Trzaska[20] at current levels of several kiloamperes. This switch, illustrated in Fig. 11, involves a rotating copper disc containing an insulating wedge-shaped segment. Current passes from one set of brushes through the disc to the other set of brushes. When the insulating segment passes under the brushes, current interruption begins. When the insulating segment leaves the brush area, continuity is restored. A drum type rotor can also be used in place of the disc. This switch has the advantages of being mechanically simple and quite compact. One disadvantage is that the entire launcher system must run at the rate set by the rotating element. The rotational speed can easily be altered, but not without altering the duration of the interruption cycle. A major hurdle faced by designers in enlarging this device is the dissipation of the stray switching energies involved with megampere current levels. As shown by Woodson and Weldon[21], a finite amount of energy must be absorbed by the switching element during each operation.*

---

* Editors' Note: Several mechanical opening switches, specifically aimed at electromagnetic launcher applications, are currently being developed by various industrial companies in the USA. Due to the proprietary nature of this work, we are unable to include any detailed discussions of these switch concepts in this book.

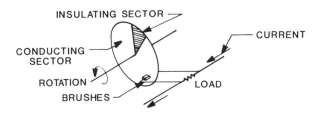

Fig. 11. Rotating mechanical switch.

DC INTERRUPTERS

There are two basic ways a mechanical switch can interrupt direct current. One is by creating an arc where the arc voltage is greater than the source voltage (often obtained by arc stretching or arc splitting). This method is commonly used in low voltage (< 5 kV) systems. The other method is to create an artificial current zero in the localized area of the switch. This method is accomplished by injecting an equal but opposite current (i.e. a counterpulse) through the switch while the contacts are in an arcing mode. The process is commonly referred to as commutation and is used in higher voltage systems.

Although a commutated high voltage switch has been proposed for HVDC transmission lines, the most common application has been for fusion experiments. One of these experiments, the JET project in Europe, utilizes a commutated air blast circuit breaker. ALCATOR, TFTR, DOUBLET III in the USA, and JT-60 in Japan all utilize commutated vacuum interrupters. A picture of a Toshiba vacuum interrupter and actuator, which was tested by this author for possible use in TFTR, is shown in Fig. 12. During this test the two interrupters were connected in series and operated at 25 kA and 25 kV. Over 1000 interruptions were performed with no failures or restrikes. A single interrupter was tested up to 45 kA at 33 kV to generate the probability of interruption curve in Fig. 13. Four similar combined interrupters (4 parallel x 2 series) serve[22] as the ohmic heating coil interrupter for JT-60 to interrupt a total current of 130 kA at 44 kV dc.

For HVDC applications, General Electric (GE) has reported interrupting 5 kA at 102 kV with an array of four series connected vacuum interrupters[23]. This approach, however, has not been readily commercialized because of the large number of series connected devices needed for operation at HVDC levels. Even with the GE high-voltage PV08 interrupter, eight series interrupters are estimated[24] to be required for operation at 400 kV.

MECHANICAL SWITCHES

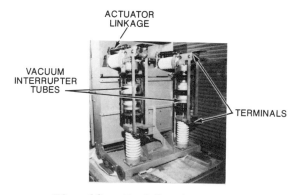

Fig. 12. Toshiba interrupter.

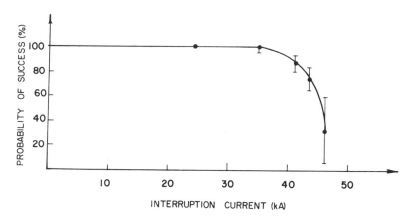

Fig. 13. Probability of interruption curve for single Toshiba interrupter.

REPETITIVE OPERATION

When considering repetitive operation, the following general conclusions can be made:

1. Almost any high current mechanical switch can be made to operate repetitively somewhere in the range of a few hertz to a few tens of hertz.

Editors' Note: Several of the basic issues facing the development of high current (MA), repetitive (1-10 pps), opening switches were addressed in a recent DoD Workshop (Tamarron VI) organized by A.K. Hyder (Auburn University) and M. Kristiansen (Texas Tech University). The proceedings of this workshop were published too late to be summarized in this chapter.

2. Obviously, the mechanism for actuating the contacts would require modification. Other considerations, such as energy dissipation schemes due to repetitive operation, actuator degradation due to increased vibrational stresses, and contact erosion and lifetime, would have to be addressed.

Let us examine the potential for repetitive operation of each typed discussed in this chapter.

  a. Pringle switch - Due to the massive size of the armature a repetition rate of only 1/2 to 1 Hz might be achieved with little modification.

  b. Brown Boveri switch - The rocker contact arrangement of this switch would permit operation at a few tens of Hertz with a pneumatic or hydraulic actuator.

  c. Wildi homopolar closing switch - This switch could operate at several tens of Hertz with moderate changes in the actuator/tripping mechanism.

  d. High-speed electro-repulsive switches possess very short closing times and could be operated repetitively if a mechanism was designed that could reload the replaceable element before each operation. The repetition rate would be governed by the rate at which these elements could be replaced since the mechanical actuating time is short.

  e. Air break switches - Although these switches open in tens of milliseconds, closing times are often as long as 1 sec. Unmodified repetition rates of 1 to 3 Hz would be possible. If the closing mechanism was altered, rates of 10 to 20 Hz might be achievable with extensive modification.

  f. Oil switches - In utility service, these switches sometimes operate several times per second when reclosing on a persistent fault. Actuator modifications could possible improve this rate to 10 Hz.

  g. Air/SF_6 blast - These switches have complex valving and linkage assemblies that would be extremely difficult to modify. In addition, the compressed gas requirement would also limit repetitive operation. Still, operation at several Hertz should be feasible. The puffer could also possibly operate at several Hertz.

  h. Vacuum switches - The inherent simplicity of these short-stroke mechanisms makes modification for repetitive operation quite reasonable. This author concluded that a repetition rate of 25 Hz could be a practical design level for such a switch[25].

SUMMARY AND CONCLUSIONS

A number of mechanical switches, both opening and closing, have been reviewed in this chapter. Both commercial and experimental types have been considered according to current and voltage ratings, opening and closing speeds and ability to perform under di/dt and dv/dt switching parameters.

REFERENCES

1. E.M. Honig and R.W. Warren, The Use of Vacuum Interrupters and By-pass Switches to Carry Currents for Long Times, Proc. 13th Pulse Power Modulator Symposium, Buffalo, NY (1978).
2. W.M. Parsons, Switching Tests for the LCTF Protective Dump Circuit, Los Alamos National Laboratory Report LA-9261-MS (1982).
3. S&C Electric Company Engineering Catalog, Descriptive Bulletin 715-31 (1981).
4. P. Wildi, High Current Making Switch, Proc. 6th Symp. on Eng. Prob. of Fus. Res., San Diego, CA (1975).
5. C.A. Bleys, D. Lebely, and F. Rioux-Damidav, A Simple Fast-closing, Metallic Contact Switch for High Voltage and Current, Rev. Sci. Instrum. 46:180 (1975).
6. P. Wildi, A Fast Metallic Contact Closing Switch for the FDX Experiment, Seminar on Energy Storage, Compression, and Switching, Canberra, Australia (1977).
7. Westinghouse Manual, "DC Breaker Application," Westinghouse Electric Corp., Switchgear Division, East Pittsburg, PA (1979).
8. General Electric, "Power Circuit Breaker ," Manual, GEH-1803B, General Electric Company, Switchgear Business Dept., Philadelphia, PA (1959).
9. Gould ITE (now Brown Boveri Electric, Inc.), "Power Circuit Breakers - DC Low Voltage," Manual IB-16.4.1.7-1 Issue B, Gould Inc., Switchgear Division, Fort Washington, PA (undated).
10. BBC, "The UR26 High Speed DC Circuit Breaker," Manual CH-SB-SHR 262E, Brown Boveri Corp., Geneva, Switzerland (undated).
11. P.L. Hartsock, T.J. Scully, and V.N. Stewart, Field Testing DC Air Circuit Breakers on a High Capacity System, Proc. AIEE Winter Power Meeting, New York, NY (1958).

12. C.E. Flurscheim, editor, "Power Circuit Breaker Theory and Design," Peter Pereginus Ltd., Southgate House, Stevenage, Herts. SG1 1HQ, England (1975).
13. T.R. Burkes, M. Kristiansen, J.P. Craig, and W.M. Portnoy, A Critical Analysis and Assessment of High Power Switches, Report NP30/78, Naval Surface Weapons Center, Special Applications Branch, Dahlgren VA (1978).
14. W.B. Whitney and E.B. Wedmore, British Patent 278764 (1926).
15. R.W Sorensen and H.E. Mendenhall, Vacuum Switching Experiments at California Institute of Technology, Trans. AIEE 45:1102 (1926).
16. Y. Sunada, N. Ito, S. Yanabu, H. Awaji, H. Okumura, and Y. Kanai, Research and Development on 13.8 kV 100 kA Vacuum Circuit-Breaker with Huge Capacity and Frequent Operation, Proc. International Conference on Large High Voltage Elect. Sys., Paris, France (1982).
17. D.R. Kurtz, J.C. Sofianek, and D.W. Crouch, Vacuum Interrupters for High Voltage Transmission Circuit Breakers, IEEE Trans. Power Apparatus and Syst. PAS-94:1094 (1975).
18. T.H. Lee, "Physics and Engineering of High Power Switching Devices," MIT Press, Cambridge, MA (1975).
19. W.M. Parsons, Design and Testing of a Prototype Water-Cooled Vacuum Interrupter for Use in Superconducting Magnet Protection Circuits, Los Alamos National Laboratory informal report LA-9687-MS (1983).
20. J.P. Barber and T.J. Trzaska, Repetitive Switching for Inductive Energy Storage, Air Force Wright Aeronautical Laboratory Report AFWAL-TR-82-2088 (1982).
21. H.H. Woodson and W.F. Weldon, Energy Considerations in Switching Current from an Inductive Store into a Raligun, Proc. 4th IEEE International Pulsed Power Conference, Albuquerque, NM (1983).
22. S. Tamura, R. Shimada, Y. Kito, Y. Kanai, H. Koike, H. Ikeda, and S. Yanabu, Parallel Interruption of Heavy Direct Current by Vacuum Circuit Breakers, IEEE Trans. on Power Apparatus and Systems PAS-99:1119 (1980).
23. A.N. Greenwood, P. Barkan, and W.C. Kracht, HVDC Vacuum Circut Breaker, IEEE Trans. Power Apparatus and Systems PAS-91:1575 (1972).
24. General Electric, Development of a HVDC Prototype Breaker, Department of Energy Report SRD-78-149, General Electric Corporate Research and Development, Schenectady, NY (1978).
25. W.M. Parsons, A Comparison Between a SCR and a Vacuum Interrupter System for Repetitive Opening, Proc. Department of Defense Workshop on Repetitive Opening Switches DTIC No. AD-A110770, (M. Kristiansen and K. Schoenbach, eds.) Tamarron, CO (1981).

CONTRIBUTORS

R.J. Commisso
Plasma Technology Branch
Plasma Physics Division
Naval Research Laboratory
Washington, DC 20375-5000

G. Cooperstein
Plasma Technology Branch
Plasma Physics Division
Naval Research Laboratory
Washington, DC 20375-5000

Laszlo J. Demeter
Physics International Co.
2700 Merced Street
San Leaandro, CA 94577-0703

R.N. DeWitt
Naval Surface Weapons Center
Code F-12
Dahlgren, VA 22448

Robin J. Harvey
Plasma Physics Department
Hughes Research Laboratories
Malibu, CA 90265

Emanuel M. Honig
Los Alamos National Laboratory
WDP/AWT, MS F668
Los Alamos, NM 87545

A.K. Jonscher
Physics Department
Royal Holloway and
   Bedford New College
Egham, Surrey, TW20, OEX, UK

Manfred Kahn
Naval Research Laboratory
Code 6360 Ceramics Branch
Washington, DC 20375

R.A. Meger
Plasma Technology Branch
Plasma Physics Division
Naval Research Laboratory
Washington, DC 20375-5000

J.M. Neri
Plasma Technology Branch
Plasma Physics Division
Naval Research Laboratory
Washington, DC 20375-5000

W.M. Parsons
Los Alamos National Laboratory
MS-D 464
P.O. Box 1663
Los Alamos, NM 87544

P.F. Ottinger
Plasma Technology Branch
Plasma Physics Division
Naval Research Laboratory
Washington, DC 20375-5000

Robert E. Reinovsky
Los Alamos National Laboratory
Group M-6, MS J970
Los Alamos, NM 87545

Gerhard Schaefer
Polytechic University
Route 110
Farmingdale, NY 11735

Karl H. Schoenbach
Dept. of Electrical Engineering
Old Dominion University
Norfolk, VA 23508

Robert W. Schumacher
Plasma Physics Department
Hughes Research Laboratories
Malibu, CA 90265

T.J. Tucker
Sandia National Laboratories
P.O. Box 5800
Albuquerque, NM 87185

P.J. Turchi
R&D Associates
Washington Research Laboratory
301 S West Street
Alexandria, VA 22314

B.N. Turman
Sandia National Laboratories
Org. 4252
P.O. Box 5800
Albuquerque, NM 87185

B.V. Weber
JAYCOR
Vienna, VA 22180

INDEX

Abnormal glow, 101
Air blast, 296-299, 302, 304
Air break switches, 293-295, 304
Anode sheath, 51, 180-181
Applications, 2, 4-6, 8, 12, 23-25, 29-30, 36-37, 44, 47-48, 50-51, 61-62, 93, 106, 110, 112, 121-122, 126-127, 133-134, 138, 149, 151, 154-155, 168, 172, 174, 195, 209, 214-215, 219, 221, 229-230, 244, 249, 259-260, 266, 274-275, 284-285, 289, 291, 299, 301-302, 306
Attacher, 51-52, 55-58, 61, 64-66, 68, 79
Attachment, 49, 51-58, 60-71, 76, 79, 82, 133-134, 138, 222, 263

Bernoulli equation, 138
Bipolar flow, 164, 180, 182, 185-186
Bipolar mode, 181-182
Bipolar space-charge, 157, 159, 163
Boltzmann distribution, 67-68
Boltzmann equation, 54-55, 138
Brown Boveri Switch, 289-290, 294-295, 300, 304-305
Bulk switching materials, 275
Burst mode, 79, 83

Carrier injection, 257, 262
Carrier production, 261
Cathode fall, 59, 99, 294
Cathode sheath, 50, 98-99, 102-

Cathode sheath, (continued) 103, 126, 156
Centered-expansion fan, 198, 200, 203
Child-Langmuir law, 99, 160, 162
Commutation, 48, 74, 93, 302
Compression wave, 203, 248
Conductance, 52, 145
Continuity equation, 133, 138-139
Control grid, 111-113, 116, 122, 126-127
Counter pulse, 302
Cross sections, 54-55, 62-63, 66-67, 70, 211, 219, 275
CROSSATRON modulator switch, 103, 109-112, 114, 116, 118, 124, 126,
 (CMS), 94, 109-111, 113, 116-117, 122, 126-127
Cross-field tube, 94, 102-104, 106, 124, 126
 (XFT), 94, 103-104, 106-109, 111, 116, 124-126
Current density, 35, 55, 58-60, 65, 72, 76-77, 79, 83, 94, 99-103, 106, 109-110, 112, 115-116, 121-122, 124-125, 127, 133-134, 142-143, 157, 159-160, 181, 192-193, 201, 215, 221, 233, 244-245, 247, 249, 254, 266, 276-277, 279, 281
Current gain, 55-56, 78-79, 84, 94, 99-103, 106, 109-110, 112, 115-116, 121-122, 124-125, 127

DC interrupters, 302

Debye, 156-157
Decay time, 96, 166-167
di/dt, 11, 29, 34-36, 41-42, 97, 216, 296, 305
Dielectric breakdown, 237-238, 257-259
Dielectric loss, 270
Dielectric strength, 54, 131-132, 146-147, 228-229, 260, 295, 298, 300
Diffuse discharge, 8, 44, 49-50, 52, 55-56, 59-62, 64, 68, 72-74, 76-79, 83-84, 93-94, 101, 145-147, 196
Diffuse mode, 25, 29, 60
Diffusion coefficient, 135, 139, 263
Dissociation, 65, 69, 71, 270
Dissociative attachment, 51, 62-63, 65-67, 71
Distribution function, 54-55
Drift velocity, 54, 56, 76, 79, 82, 133, 161, 194, 262
Duty factor, 4
dv/dt, 11, 24, 29, 34-36, 41-42, 299, 305
Dynamic quenches, 229

Effective resistivity, 215-216, 219-220, 276
Efficiency, 12-13, 15-16, 42, 50, 55, 74, 77-78, 84, 100, 107, 146, 151, 167, 185, 204, 247-249, 275, 283
Electrical aging, 258
Electron beam (e-beam), 8, 49-50, 52, 54-61, 69, 72-73, 83-84, 151, 154, 156, 164, 169, 173-174, 177, 181, 187-189
Electronegative, 53, 60
Energy capacitive, 2-3, 5-6, 12, 16-19, 21-23, 40, 42, 149, 175
Energy dissipation, 16, 23, 216
Energy inductive, 1-5, 7-10, 12-16, 19, 21-23, 25, 38, 41-45, 49-51, 54, 74, 74, 79, 83, 107-108, 110, 119, 149, 151, 155, 168-

Energy inductive (continued), 173, 175, 177, 183, 185, 188, 191, 194, 202, 204, 211-212, 216, 222, 226, 233, 245, 247, 306
Energy inertial, 5-6, 8, 169, 228
Energy ratio, 14, 15, 20, 21
Energy storage, 1-4, 6-7, 9, 16, 23, 25, 38, 44, 45, 52, 107-108, 110, 149, 171-172, 177-178, 187-188, 245, 273, 288, 305-306
Equation of state, 211, 219
Excitation, 50, 61, 64, 66-69, 71-72, 100
Excited states, 54, 64, 69-71
Expansion wave, 203
Explosive product, 233-235, 237, 240, 242, 248-249, 252, 254
Explosively-driven opening switches, 233
Fault current, 2, 296, 298
Field emission, 72, 96, 102, 105, 157, 174
Flyer motion, 241
Foil fuses, 218-220, 223, 226, 228-229
Foil material, 74
Fowler-Nordheim process, 269
Frozen-field, 195, 197, 200, 202, 204, 207
Fuse length, 213, 218, 229
Fuse material, 210-214, 216, 218-219
Fuse modeling, 214
Fuse opening switch, 209, 230
Fuse performance, 220-222, 230
Fuses in gaseous media, 228
Fuses in liquid media, 228
Fuses in solid media, 223

Gamble I generator, 151-152, 167, 176
Gas engineering, 84
Glow discharge, 61, 93-95, 97, 100-102, 105, 147
Glow-to-arc transition, 60, 76, 102, 106
Grain boundary, 275-278

INDEX

Grid potential, 115

Homopolar generator, 150, 171, 291

Instabilities, 49, 51, 59-60, 74, 100, 204, 219, 229
Instability, 60, 242-243
Intrinsic breakdown
Ionization, 46, 49, 51, 54-55, 59-61, 93, 95-100, 106, 108, 112-113, 115-116, 133-136, 141, 143, 147, 204, 222-223, 228, 258, 262-263, 268, 282
Ionization energy, 50, 55, 100, 136
Ionization source, 52, 61, 72, 108, 112

Langmuir, 99, 115, 179-181, 185
Langmuir-Child current, 179
Larmor radius, 158, 163, 165, 167
Lifetime, 64, 126-127, 157, 304
Load, 1-5, 8-23, 37-40, 42-43, 49, 52, 74, 78, 83, 107, 150-155, 157, 160-163, 166-171, 173-174, 177-178, 191, 196-198, 200, 204, 206-207, 212-213, 215, 217, 227, 233-234, 249-251, 273-274, 292
Load line, 83

Magnetic field, 29-30, 46, 93-94, 96, 99, 102-104, 106-107, 109, 111-113, 115-116, 124-125, 132, 149, 156, 160, 162-163, 165, 177-178, 192-203, 205-207, 219, 221, 275, 277-278, 283, 294, 300
Magnetic insulation, 157, 162, 164, 167, 175, 207
Magnetic pressure, 194, 203, 219
Magnetic Reynold's number, 195-197, 207
Magneto-Hydro-Dynamic (MHD), 218, 221, 241

Magnetoresistive switching, 278, 283
Maxwellian distribution, 142
Mechanical switches, 287-288, 291, 305
Mobility, 51-52, 139, 192, 262-263, 265, 267, 273, 275, 277
Monte Carlo, 54-55, 147
Multi-photon dissociation, 69, 71

Negative differential conductivity, 56-57, 71, 79, 83
Nernst's equation, 136

Ohm's law, 192-193, 195
Oil switches, 295, 304
Opening switches, 1-4, 8-13, 16-17, 19, 22, 25, 30, 36, 38, 41, 43-45, 48-49, 51, 61-62, 64, 74, 76, 79, 83, 93-94, 96, 102, 110, 120, 123, 133, 147, 149, 155, 171, 173-174, 178-179, 191, 194, 196-197, 200, 204-205, 209-210, 230, 233-334, 238, 240, 249-251, 254, 273-274, 278, 287-288, 293-294, 301, 303, 306
 current-zero, 8-10, 16-17, 22-25, 29-30, 33-34, 36-37, 39-42
 direct-interruption, 8-9, 12, 16, 22-23, 34
Optical control, 50, 54, 58, 61-62
Optically enhanced attachment, 64, 71

Partial discharges, 262
Paschen breakdown curve, 94-95, 97, 109
Phase transformation, 274
Photo detachment, 58
Planar switch, 240, 242
Plasma convection, 197-198, 206
Plasma erosion, 149-150, 178, 196, 207

Plasma erosion opening switch (PEOS), 149-156, 160, 164-165, 167-175,
Plasma flow, 191, 196-197, 200, 202, 204-206
Plasma flow switch, 191, 196-197, 204-206
Plasma injection, 70, 102, 109, 153, 155, 157, 159, 165
Plasma potential, 111, 114-115
Plasma sources, 165, 169, 171, 202
Plasmadynamic opening switch, 191, 204
Poisson's equation, 142, 192
Poynting vector, 192-194, 196, 202
Precursor, 198, 202-204
Pringle switch, 289, 304
Pulse sharpening, 2-3, 173
Pulse width, 4, 78, 149, 171

Quench material, 217

Raether's criterium, 132
Rate coefficients, 51-52, 54, 57, 60, 62, 71
Rate equations, 51-52, 54
Rayleigh-Taylor instability, 242
Recombination, 33, 49, 51-53, 56-58, 76, 133, 138, 274
Recovery, 9, 11, 16, 23-25, 29-30, 33-36, 41, 93, 96, 110, 131-133, 135, 137, 140, 143-147, 210-211, 276, 296, 298-300
Reduced electric field (E/N), 135, 143
Reflex switch, 177-179, 181-183, 187
Reflex triode, 178-181, 185-186
Relativistic electron beam (REB), 174, 177, 185, 187
Repetition rate, 4, 6, 23, 25, 37, 38, 93, 131, 133, 135, 143, 146, 273, 288, 290, 304-305
Repetitive operation, 74, 132, 273, 287-289, 292, 300, 303-304

Rise time, 9, 17-18, 22, 34, 37-42, 78, 233, 254
Rocker switch, 291
Rod array triggered vacuum gap (see also RATVG), 25-26, 28
Saha equation, 135-137
Saturable reactor, 11, 35, 41
SCR, 30-35, 47, 306
Secondary yield, 100, 121
SF_6 gas blast, 293, 298
Solid state opening switch, 273
Source function, 52, 55, 57
Source grid, 111-113
Space charge, 99, 112, 115, 141-143, 160, 162-163, 174, 180, 276
Superconducting materials, 7
Sustained discharge, 49-51, 54-55, 57-58, 61, 64, 69, 71, 74, 76, 79, 83-84, 134-135
Stray inductance, 2-3, 12, 14, 16, 18, 34
Streamer, 60, 132, 134-135
Switching time, 167, 284

Tamper, 223, 226, 337, 240-241
Thermal aging
Thermal breakdown, 265
Thermionic emitters, 74
Thyristor, 10-11, 23, 30-31, 33-36, 41-42
Time constant, 8-9, 14, 29-30, 133, 144, 240
Townsend avalanche, 132, 134, 144
Transfer capacitor, 16-20, 23
Transfer circuits, 12-13, 16, 19, 22, 26
Transient behavior, 50, 64
Turn-off, 30-31, 33-36, 47
Turn-on, 25, 32-36, 41-42, 186

UV, 54, 61, 69-71, 204

Vacuum, 10, 23-27, 29-31, 38, 41-42, 45-46, 72, 93, 96-97, 102, 105, 107, 109-111, 113, 123-124, 126-127, 131, 143, 150-151, 171, 173-174, 177-179, 183, 187, 197-204, 206, 209,

Vacuum (continued), 219, 228-229, 259, 273, 293, 299-302, 304-306
Vacuum circuit breaker, 299, 306

Vacuum interrupter, 23-25, 27, 29, 38, 42, 45-46, 294-296, 298-303, 305-306
Vacuum switches, 10, 23-27, 29-31, 41, 46, 133, 147-148, 177, 187, 299, 304
Vaporization, 198, 204, 209-219, 221-222, 228-229
Varistor, 34, 41, 119, 284
Vibrational excitation, 61, 64, 66-71
Vibrational states, 66-69, 71, 143
Voltage recovery, 29, 96, 144

Weibull statistics, 259-260, 263, 265
Wire-ion-plasma electron gun (WIP), 73, 105